A WALK FROM LONDON TO JOHN O'GROAT

Publisher's Note

Purchase of this book entitles you to a free trial membership in the publisher's book club at www.rarebooksclub.com. (Time limited offer.) Simply enter the barcode number from the back cover onto the membership form on our home page. The book club entitles you to select from millions of books at no additional charge. You can also download a digital copy of this and related books to read on the go. Simply enter the title or subject onto the search form to find them.

Note: This is an historic book. Pages numbers, where present in the text, refer to the first edition of the book. The table of contents or index may also refer to them.

If you have any questions, could you please be so kind as to consult our Frequently Asked Questions page at www.rarebooksclub.com/faqs.cfm? You are also welcome to contact us there.

Publisher: General Books LLC™, Memphis, TN, USA, 2012. ISBN: 9781443215985.

Credits: Distributed Proofreaders.

✂ ✂ ✂ ✂ ✂ ✂ ✂ ✂

CONTENTS.

PREFACE.

In presenting this volume to the public,

I feel that a few words of explanation are due to the readers that it may obtain, in addition to those offered to them in the first chapter. When I first visited England, in 1846, it was my intention to make a pedestrian tour from one end of the island to the other, in order to become more acquainted with the country and people than I could by any other mode of travelling. A few weeks after my arrival, I set out on such a walk, and had made about one hundred miles on foot, when I was constrained to suspend the tour, in order to take part in movements which soon absorbed all my time and strength. For the ensuing ten years I was nearly the whole time in Great Britain, travelling from one end of the kingdom to the other, to promote the movements referred to; still desiring to accomplish the walk originally proposed. On returning to England at the beginning of 1863, after a continuous residence of seven years in America, I found myself, for the first time, in the condition to carry out my intention of 1846. Several new motives had been added in the interval to those that had at first operated upon my mind. I had dabbled a little in farming in my native village, New Britain, Connecticut, and had labored to excite additional interest in agriculture among my neighbors. We had formed an Agricultural Club, and met weekly for several winters to compare notes, exchange opinions' and discuss matters connected with the occupation. They had honored me with the post of Corresponding Secretary from the beginning. We held a meeting the evening before I left for England, when they not only refused to accept my resignation as Secretary, but made me promise to write them letters about farming in the Mother Country, and on other matters of interest that I might meet with on my travels there. My first idea was to do this literally;—to make a walk through the best agricultural sections of England, and write home a series of communications to be inserted in our little village paper. But, on second thought, on considering the size of the sheet, I found it would require four or five years to print in it all I was likely to write, at the rate of two columns a week. So I concluded that the easiest and quickest way would be to make a book of my Notes by the Way, and to send back to my old friends and neighbors in that form all the observations and incidents I might make and meet on my walk. The next thought that suggested itself was this,—that a good many persons in Great Britain might feel some interest in seeing what an American, who had resided so long in this country, might have to say of its sceneries, industries, social life, etc. Still, in writing out these Notes, although two distinct circles of readers—the English and American—have been present to my mind, I felt constrained to face and address the latter, just as if speaking to them alone. I have, moreover, adopted the free and easy style of epistolary composition, endeavoring to make each chapter as much like one of the letters I promised my friends and neighbors at home as practicable. In doing this, the "I" has, perhaps, talked far too much to beseem those proprieties which the author of a book should observe. Besides, expressions, figures and orthography more American than English may be noticed, which will indicate the circle of readers which the writer had primarily in view. Still, he would fain believe that these features of the volume will not seriously affect the interest it might otherwise possess in the minds of those disposed to give it a reading in this country. Whatever exceptions they may take to the style and diction, I hope they will find none to the spirit of the work.

ELIHU BURRITT.

London, April 5th, 1864.

CHAPTER I.

MOTIVES TO THE WALK—THE IRON HORSE AND HIS RIDER—THE LOSSES AND GAINS BY SPEED—THE RAILWAY TRACK AND TURNPIKE ROAD: THEIR SCENERIES COMPARED.

Elihu Burritt

One of my motives for making this tour was to look at the country towns and villages on the way in the face and eyes; to enter them by the front door, and to see them as they were made to be seen first, as far as man's mind and hand intended and wrought. Railway travelling, as yet, takes everything at a disadvantage; it does not front on nature, or art, or the common conditions and industries of men in town or country. If it does not actually of itself turn, it presents everything the wrong side outward. In cities, it reveals the ragged and smutty companionship of tumble-down out-houses, and mysteries of cellar and back-kitchen life which were never intended for other eyes than those that grope in them by day or night. How unnatural, and, more, almost profane and inhuman, is the fiery locomotion of the Iron Horse through these densely-peopled towns! now the screech, the roar, and the darkness of cavernous passages under paved streets, church vaults, and an acre or two of three-story brick houses, with the feeling of a world of breathing, bustling humanity incumbent upon you;—now the dash and flash out into the light, and the higgledy-piggledy glimpses of the next five minutes. In a moment you are above thickly-thronged streets, and the houses on either side, looking down into the black throats of smoky chimneys; into the garret lairs of poverty, sickness, and sin; down lower upon squads of children trying to play in back-yards eight feet square. It is all wrong, except in the single quality of speed. You enter the town as you would a farmer's house, if you first passed through the pig-stye into the kitchen. Every respectable house in the city turns its back upon you; and often a very brick and dirty back too, though it may show an elegant front of Bath or Portland stone to the street it faces. All the respectable streets run over or under you with an audible shudder of disgust or dread. None but a shabby lane of low shops for the sale of junk, beer, onions, shrimps, and cabbages, will run a third of a mile by your side for the sake of your company. The wickedest boys in the town hoot at you, with most ignominious and satiric antics, as you pass; and if they do not shie stones in upon you, or dead cats, it is more from fear of the beadle or the constable than out of respect for your business or pleasure.

Indeed, every town and village, great or small, which you pass through or near on the railway, looks as if you came fifty years before you were expected. It says, in all the legible expressions of its countenance, "Lack-a-day!—if here isn't that creature come already, and looking in at my back door before I had time to turn around, or put anything in shape!" The Iron Horse himself gets no sympathy nor humane admiration. He stands grim and wrathy, when reined up for two minutes and forty-five seconds at a station. No venturesome boys pat him on the flanks, or look kindly into his eyes, or say a pleasant word to him, or even wonder if he is tired, or thirsty, or hungry. None of the ostlers of the greasy stables, in which the locomotives are housed, ever call him Dobbin, or Old Jack, or Jenny, or say, "Well done, old fellow!" when they unhitch him from the train at midnight, after a journey of a hundred leagues. His driver is a real man of flesh and blood; with wife and children whom he loves. He goes on Sunday to church, and, maybe, sings the psalms of David, and listens devoutly to the sermon, and says prayers at home, and the few who know him speak well of him, as a good

and proper man in his way. But, spurred and mounted upon the saddle of the great iron hexiped, nearly all the passengers regard him as a part of the beast. No one speaks to him, or thinks of him on the journey. He may pull up at fifty stations, and not a soul among the Firsts, Seconds, or even Thirds, will offer him a glass of beer, or pipe-full of tobacco, or give him a sixpence at the end of the ride for extra speed or care. His face is grimy, and greasy, and black. All his motions are ambiguous and awkward to the casual observer. He has none of the sedate and conscious dignity of his predecessor on the old stage-coach box. He handles no whip, like him, with easy grace. Indeed, in putting up his great beast to its best speed, he "hides his whip in the manger," according to a proverb older than steam power. He wears no gloves in the coldest weather; not always a coat, and never a decent one, at his work. He blows no cheery music out of a brass bugle as he approaches a town, but pricks the loins of the fiery beast, and makes him scream with a sound between a human whistle and an alligator's croak. He never pulls up abreast of the station-house door, in the fashion of the old coach driver, to show off himself and his leaders, but runs on several rods ahead of his passengers and spectators, as if to be clear of them and their comments, good or bad. At the end of the journey, be it at midnight or daybreak, not a man nor a woman he has driven safely at the rate of forty miles an hour thinks or cares what becomes of him, or separates him in thought from the great iron monster he mounts. Not the smock-frocked man, getting out of the forwardmost Third, with his stick and bundle, thinks of him, or stops a moment to see him back out and turn into the stable.

With all the practical advantages of this machine propulsion at bird speed over space, it confounds and swallows up the poetical aspects and picturesque sceneries that were the charm of old-fashioned travelling in the country. The most beautiful landscapes rotate around a locomotive axis confusedly. Green

pastures and yellow wheat fields are in a whirl. Tall and venerable trees get into the wake of the same motion, and the large, pied cows ruminating in their shade, seem to lie on the revolving arc of an indefinite circle. The views dissolve before their best aspect is caught by the eye. The flowers, like Eastern beauties, can only be seen "half hidden and half revealed," in the general unsteadiness. As for bees, you cannot hear or see them at all; and the songs of the happiest birds are drowned altogether by the clatter of a hundred wheels on the metal track. If there are any poor, flat, or fen lands, your way is sure to lie through them. In a picturesque and undulating country, studded with parks and mansions of wealth and taste, you are plunging through a long, dark tunnel, or walled into a deep cut, before your eye can catch the view that dashes by your carriage window. If you have a utilitarian proclivity and purpose, and would like to see the great agricultural industries of the country, they present themselves to you in as confused aspects as the sceneries of the passing landscape. The face of every farm is turned from you. The farmer's house fronts on the turnpike road, and the best views of his homestead, of his industry, prosperity, and happiness, look that way. You only get a furtive glance, a kind of clandestine and diagonal peep at him and his doings; and having thus travelled a hundred miles through a fertile country you can form no approximate or satisfactory idea of its character and productions.

But no facts nor arguments are needed to convince an intelligent traveller that the railway affords no point of view for seeing town or country to any satisfactory perception of its character. Indeed, neither coach of the olden, nor cab of the modern vogue, nor saddle, will enable one to "do" either town or country with thorough insight and enjoyment. It takes him too long to pull up to catch the features of a sudden view. He can do nothing with those generous and delightful institutions of Old England,—the footpaths, that thread pasture, park, and field, seemingly permeating her whole green world with dusky veins for the circulation of human life. To lose all the picturesque lanes and landscapes which these field-paths cross and command, is to lose the great distinctive charm of the country. Then, neither from the coach-box nor the saddle can he make much conversation on the way. He loses the chance of a thousand little talks and pleasant incidents. He cannot say "Good morning" to the farmer at the stile, nor a word of greeting to the reapers over the hedge, nor see where they live, and the kind of children that play by their cottage doors; nor the little, antique churches, bearded to their eye-brows with ivy, covering the wrinkles of half a dozen centuries, nor the low and quiet villages clustering around, each like a family of bushy-headed children surrounding their venerable mother.

In addition to these considerations, there was another that moved me to this walk. Although I had been up and down the country as often and as extensively as any American, perhaps, and admired its general scenery, I had never looked at it with an agricultural eye or interest. But, having dabbled a little in farming in the interval between my last two visits to England, and being touched with some of the enthusiasm that modern novices carry into the occupation, I was determined to look at the agriculture of Great Britain more leisurely and attentively, and from a better stand-point than I had ever done before. The thought had also occurred to me, that a walk through the best agricultural counties of England and Scotland would afford opportunity for observation which might be made of some interest to my friends and neighbor farmers in America as well as to myself. Therefore I beg the English reader to remember that I am addressing to them the notes that I may make by the way, hoping that its incidents and the thoughts it suggests will not be devoid of interest because they are principally intended for the American ear.

CHAPTER II.

FIRST DAY'S OBSERVATIONS AND ENJOYMENT—RURAL FOOT-PATHS; VISIT TO TIPTREE FARM—ALDERMAN MECHI'S OPERATIONS—IMPROVEMENTS INTRODUCED, DECRIED, AND ADOPTED—STEAM POWER, UNDER-DRAINING, DEEP TILLAGE, IRRIGATION—PRACTICAL RESULTS.

On Wednesday, July 15th, 1863, I left London with the hope that I might be able to accomplish the northern half of my proposed "Walk from Land's End to John O'Groat's." I had been practically prostrated by a serious indisposition for nearly two months, and was just able to walk one or two miles at a time about the city. Believing that country air and exercise would soon enable me to be longer on my feet, I concluded to set out as I was, without waiting for additional strength, so slow and difficult to attain in the smoky atmosphere and hot streets of London.

Few reading farmers in America there are who are not familiar with the name and fame of Alderman Mechi, as an agriculturist of that new and scientific school that is making such a revolution in the great primeval industry of mankind. His experiments on his Tiptree Farm have attained a world-wide publicity, and have given that homestead an interest that, perhaps, never attached to the same number of acres in any country or age. Thinking that this famous establishment would be a good starting point for my pedestrian tour, I concluded to proceed thither first by railway, and thence to walk northward, by easy stages, through the fertile and rural county of Essex. Taking an afternoon train, I reached Kelvedon about 5 p.m.,—the station for Tiptree, and a good specimen of an English village, at two hours' ride from London. Calling at the residence of a Friend, or Quaker, to inquire the way to the Alderman's farm, he invited me to take tea with him, and be his guest for the night,—a hospitality which I very gladly accepted, as it was a longer walk than I had antic-

ipated. After tea, my host, who was a farmer as well as miller, took me over his fields, and showed me his live stock, his crops of wheat, barley, oats, beans, and roots, which were all large and luxuriant, and looked a *tableau vivant* of plenty within the green hedges that enclosed and adorned them.

The next morning, after breakfast, my kind host set me on the way to Tiptree by a footpath through alternating fields of wheat, barley, oats, beans, and turnips, into which an English farm is generally divided. These footpaths are among the vested interests of the walking public throughout the United Kingdom. Most of them are centuries old. The footsteps of a dozen generations have given them the force and sanctity of a popular right. A farmer might as well undertake to barricade the turnpike road as to close one of these old paths across his best fields. So far from obstructing them, he finds it good policy to straighten and round them up, and supply them with convenient gates or stiles, so that no one shall have an excuse for trampling on his crops, or for diverging into the open field for a shorter cut to the main road. Blessings on the man who invented them! It was done when land was cheap, and public roads were few; before four wheels were first geared together for business or pleasure. They were the doing of another age; this would not have produced them. They run through all the prose, poetry, and romance of the rural life of England, permeating the history of green hedges, thatched cottages, morning songs of the lark, moonlight walks, meetings at the stile, harvest homes of long ago, and many a romantic narrative of human experience widely read in both hemispheres. They will run on for ever, carrying with them the same associations. They are the inheritance of landless millions, who have trodden them in ages past at dawn, noon, and night, to and from their labor; and in ages to come the mowers and reapers shall tread them to the morning music of the lark, and through Spring, Summer, Autumn, and Winter, they shall show the fresh checker-work of the ploughman's hob-nailed shoe. The surreptitious innovations of utilitarian science shall not poach upon these sacred preserves of the people, whatever revolutions they may produce in the machinery and speed of turnpike locomotion. These pleasant and peaceful paths through park, and pasture, meandering through the beautiful and sweet-breathing artistry of English agriculture, are guaranteed to future generations by an authority which no legislation can annul.

A walk of a few miles brought me in sight of Tiptree Hall; and its first aspect relieved my mind of an impression which, in common with thousands better informed, I had entertained in reference to the establishment. An idea has generally prevailed among English farmers, and agriculturists of other countries who have heard of Alderman Mechi's experiments, that they were impracticable and almost valueless, because they would not *pay*; that the balance-sheet of his operations did and must ever show such ruinous discrepancy between income and expenditure as must deter any man, of less capital and reckless enthusiasm, from following his lead into such unconsidered ventures. In short, he has been widely regarded at home and abroad as a bold and dashing novice in agricultural experience, ready to lavish upon his own hasty inventions a fortune acquired in his London warehouse; and all this to make himself famous as a great light in the agricultural world, which light, after all, was a mere will-o'-the-wisp sort of affair, leading its dupes into the veriest bog of bankruptcy. In common with all those bold, self-reliant spirits that have ventured to break away from the antecedents of public opinion and custom, he has been the subject of many ungenerous innuendoes and criticisms. All kinds of ambitions and motives have been ascribed to him. Many a burly, red-faced farmer, who boasts of an unbroken agricultural lineage reaching back into the reign of Good Queen Bess, will tell you over his beer that the Alderman's doings are all *gammon*; that they are all to advertise his cutlery business in Leadenhall Street, Barnum fash-

ion; to inveigle down to Tiptree Hall noblemen, foreign ambassadors, and great people of different countries, and bribe "an honourable mention" out of them with champagne treats and oyster suppers. Indeed, my Quaker host largely participated in this opinion, and took no pains to conceal it when speaking of his enterprising neighbor.

From what I had read and heard of the Tiptree Hall estate, I expected to see a grand, old, baronial mansion, surrounded with elegant and costly buildings for housing horses, cattle, sheep, and other live stock, all erected on a scale which no *bona fide* farmer could adopt or approximately imitate. In a word, I fancied his barns and stables would even surpass in this respect the establishments of some of those most wealthy New York or Boston merchants, who think they are stimulating country farmers to healthy emulation by lavishing from thirty to forty thousand dollars on a barn and its appurtenant out-houses. With these preconceived ideas, it was an unexpected satisfaction to see quite a simple-looking, unassuming establishment, which any well-to-do farmer might make and own. The house is rather a large and solid-looking building, erected by Mr. Mechi himself, but not at all ostentatious of wealth or architectural taste. The barns and "steddings," or what we call cowhouses in America, are of a very ordinary cast, or such as any country-bred farmer would call economical and simple. The homestead occupies no picturesque site, and commands no interesting scenery. The farm consists of about 170 acres, which, in England, is regarded as a rather small holding. The land is naturally sterile and hard of cultivation, most of it apparently being heavily mixed with ferruginous matter. When ploughed deeply, the clods turned up look frequently like compact masses of iron ore. Every experienced farmer knows the natural poverty of such a soil, and the hard labor to man and beast it costs to till it.

To my great regret, Mr. Mechi was not at home, though he passes most of his time in Summer at Tiptree. But his foreman, who enters into all the experi-

ments and operations which have made the establishment so famous, with almost equal interest and enthusiasm, took me through the farm buildings, and all the fields, and showed me the whole process and machinery employed. Any English or American agriculturist who has read of Alderman Mechi's operations, would be inclined to ask, on looking, for the first time, at his buildings and the fields surrounding them, what is the great distinguishing speciality of his enterprise. His land is poor; his housings are simple; there is no outside show of uncommon taste or genius. Every acre is tile-drained, to be sure. But that is nothing new nor uncommon. Drainage is the order of the day. Any tenant farmer in England can have his land drained by the Government by paying six per cent. annually on the cost of the job. His expenditure for artificial manure does not exceed that of hundreds of good farmers. He carries out the deep tillage system most liberally. So do other scientific agriculturalists in Europe and America. Of course, a few hours' observation would not suffice for a full and correct conclusion on this point, but it gave me the impression that the great operation which has won for the Tiptree Farm its special distinction is its irrigation with liquid manure. In this respect it stands unrivalled, and, perhaps, unimitated. And this, probably, is the head and front of his offending to those who criticise his economy and decry his experiments.

This irrigation is performed through the medium of a small steam engine and sixteen hydrants, so posted and supplied with hose as to reach every square foot of the 170 acres. The water used for this purpose is mostly, if not entirely, supplied from the draining pipes, even in the dryest season. The manure thus liquified is made by a comparatively small number of animals. Calves to the value of £50 are bought, and fat stock to that of £500 are sold annually. They are all stabled throughout the year, except in harvest time, when they are turned out for a few weeks to rowen feed. The calves are housed until a year old in a large stedding by themselves. They are then transferred to another building, and put upon "the boards;" that is in a long stable or cowhouse, with a flooring of slats, through which the manure drops into a cellar below, made watertight. Here the busiest little engine in the world is brought to bear upon it, with all its faculties of suction and propulsion. Through one pipe it forces fresh water in upon this mass of manure, which, when liquified, runs down into a subterranean cistern or reservoir capable of holding over 100,000 gallons. From this it is propelled into any field to be irrigated. To prevent any sediment in the great reservoir, or to make an even mixture of the liquified manure, a hose is attached to the engine, and the other end dropped into the mass. Through this a constant volume of air is propelled with such force as to set the whole boiling and foaming like a little cataract. One man at the engine and two at the hose in the distant field perform the whole operation. The chapped and "baky" surface of the farm is thus softened and enriched at will, and rendered productive.

Now, this operation seems to constitute the present distinctive speciality of Alderman Mechi's Tiptree Farm. Will it pay? ask a thousand voices. In how many years will he get his money back? Give us the balance sheet of the experiment. A New Englander, favorably impressed with the process, would be likely to answer these questions by another, and ask, will *drainage* pay? Not in one year, assuredly, nor in five; not in ten, perhaps. The British Government assumes that all the expenditure upon under-drainage will be paid back in fifteen or twenty years at the farthest. It lends money to the land-owner on this basis; and the land-owner stipulates with his tenant that he shall reimburse him by annual instalments of six or seven per cent. until the whole cost of the operation is liquidated. Thus the tenant-farmer is willing to pay six, sometimes seven per cent. annually, for twenty years, for the increased capacity of production which drainage gives to the farm he cultivates. At the end of that period the Government is paid by the land-lord, and the landlord by the tenant, and the tenant by his augmented crops for the whole original outlay upon the land. For aught either of the three parties to the operation knows to the contrary, it must all be done over again at the end of twenty years. The system is too young yet, even in England, for any one to say how long a course of tubing will last, or how often it must be relaid.

One point, therefore, has been gained. No intelligent English farmer, who has tried the system, now asks if under-drainage will pay; nor does he expect that it will pay back the whole expenditure in less than twelve or fifteen years. Here is a generous faith in the operation on the side of all the parties concerned. Then why should not Alderman Mechi's irrigation system be put on the same footing, in the matter of public confidence? It is nothing very uncommon even for a two-hundred-acre farmer in England to have a small stationary or locomotive steam-engine, and to find plenty of work for it, too, in threshing his grain, grinding his fodder, pulping his roots, cutting his hay and straw, and for other purposes. Mr. Mechi would doubtless have one for these objects alone. So its cost must not be charged to the account of irrigation. A single course of iron tubing, a third of a mile long, reaching to the centre of his farthest field, cannot cost more, with all the hose employed, than the drainage of that field, while it would be fair to assume that the iron pipes will last twice as long as those of burnt clay. They might fairly be expected to hold good for forty years. If, then, for this period, or less, the process yields ten per cent. of increased production annually, over and above the effect of all other means employed, it is quite evident that it will pay as well as drainage.

But does it augment the yearly production of the farm by this amount? To say that it is the only process by which the baky and chappy soil of Tiptree can be thoroughly fertilised, would not suffice to prove its necessity or value to other soils of different composition. One fact, however, may be sufficient to determine its virtue. The fields of clover, and Italian rye-grass, etc., are mown three and even four times in one season, and afterwards fed with sheep. Certainly, no other system could produce all this cropping. The distinctive difference it makes in other crops cannot, perhaps, be made so palpable. The wheat looked strong and heavy, with a fair promise of forty-five bushels to an acre. The oats, beans, and roots showed equally well.

The irrigation and deep tillage systems were going on simultaneously in the same field, affording me a good opportunity of seeing the operation of both. Two men were plying the hose upon a portion of the field which had already been mowed three times. Two teams were at work turning up the other, which had already been cropped once or twice. One of two horses went first, and, with a common English plough, turned an ordinary furrow. Then the other followed, of twice the force of the first, in the same furrow, with a subsoil plough held to the work beam-deep. The iron-stones and ferruginous clods turned up by this "deep tillage" would make a prairie farmer of Illinois wonder, if not shudder, at the plucky and ingenious industry which competes with his easy toil and cheap land in providing bread for the landless millions of Great Britain.

The only exceptional feature or arrangement, besides the irrigating machinery and process, that I noticed, was an iron hurdling for folding sheep. This, at first sight, might look to a practical farmer a little extravagant, indicating a city origin, or the notion of an amateur agriculturist, more ambitious of the new than of the necessary. Each length of this iron fencing is apparently about a rod, and cost £1, or nearly five dollars. It is fitted to low wheels, or rollers, on an axle two or three feet in length, so that it can be moved easily and quickly in any direction. It would cost over fifty pounds, or two hundred and fifty dollars, to enclose an acre entirely with this kind of hurdling. Still, Mr. Mechi would doubtless be able to show that this large expenditure is a good investment, and pays well in the long run. The folding of sheep for twenty-four or forty-eight hours on small patches of clover, trefoil, or turnips, is a very important department of English farming, both for fattening them for the market and for putting the land in better heart than any other fertilising process could effect. Now, a man with this iron fencing on wheels must be able to make in two hours an enclosure that would cost him a day or more of busy labor with the old wooden hurdles.

On the whole, a practical farmer, who has no other source of income than the single occupation of agriculture, would be likely to ask, what is the realised value of Alderman Mechi's operations to the common grain and stock-growers of the world? They have excited more attention or curiosity than any other experiments of the present day; but what is the real *resume* of their results? What new principles has he laid down; what new economy has he reduced to a science that may be profitably utilised by the million who get their living by farming? What has he actually done that anybody else has adopted or imitated to any tangible advantage? These are important questions; and this is the way he undertakes to answer them, beginning with the last.

About twenty years ago, he inaugurated the system of under-draining the heavy tile-clay lands in Essex. Up to his experiment, the process was deemed impracticable and worthless by the most intelligent farmers of the county. It was more confidently decried than his present irrigation system. The water would never find its way down into the drain-pipes through such clay. It stood to reason that it would do no such thing. Did not the water stand in the track of the horse's hoof in such rich clay until evaporated by the sun? It might as well leak through an earthenware basin. It was all nonsense to bury a man's money in that style. He never would see a shilling of it back again. In the face of these opinions, Mr. Mechi went on, training his pipes through field after field, deep below the surface. And the water percolated through the clay into them, until all these long veins formed a continuous and rushing stream into the main artery that now furnishes an ample supply for his stabled cattle, for his steam engine, and for all the barnyard wants. His tile-draining of clay-lands was a capital success; and those who derided and opposed it have now adopted it to their great advantage, and to the vast augmentation of the value and production of the county. Here, then, is one thing in which he has led, and others have followed to a great practical result.

His next leading was in the way of agricultural machinery. He first introduced a steam engine for farming purposes in a district containing a million of acres. That, too, at the outset, was a fantastic vagary in the opinion of thousands of solid and respectable farmers. They insisted the Iron Horse would be as dangerous in the barn-yard or rick-yard as the very dragon in Scripture; that he would set everything on fire; kill the men who had care of him; burst and blow up himself and all the buildings into the air; that all the horses, cows, and sheep would be frightened to death at the very sight of the monster, and never could be brought to lie down in peace and safety by his side, even when his blood was cold, and when he was fast asleep. To think of it! to have a tall chimney towering up over a barn-gable or barn-yard, and puffing out black coal smoke, cotton-factory-wise! Pretty talk! pretty terms to train an honest and virtuous farmer to mouth! Wouldn't it be edifying to hear him string the yarn of these new words! to hear him tell of his *engineer* and ploughman; of his *pokers* and pitchforks; of *six-horse power, valves, revolutions, stopcocks, twenty pounds of steam*, etc.; mixing up all this ridiculous stuff with yearling-calves, turnips, horse-carts, oil-cake, wool, bullocks, beans, and sheep, and other vital things and interests, which forty centuries have looked upon with reverence! To plough, thresh, cut turnips, grind corn, and pump water for cattle by steam! What next?

Why, next, the farmers of the region round about

"First pitied, then embraced"

this new and powerful auxiliary to agricultural industry, after having watched its working and its worth. And now, thanks to such bold and spirited novices as Mr. Mechi—men who had the pluck to work steadily on under the pattering rain of derisive epithets—there are already nearly as many steam engines working at farm labor between Land's End and John O'Groat's as there are employed in the manufacture of cotton in Great Britain.

His irrigation system will doubtless be followed in the same order and interval by those who have pooh-poohed it with the same derision and incredulity as the other innovations they have already adopted. The utilising of the sewage of large towns, especially of London, has now become a prominent idea and movement. Mr. Mechi's machinery and process are admirably adapted to the work of distributing a river of this fertilising material over any farm to which it may be conducted. Thus, there is good reason to believe that the very process he originated for softening and enriching the hard and sterile acres of his small farm in Essex will be adopted for saturating millions of acres in Great Britain with the millions of tons of manurial matter that have hitherto blackened and poisoned the rivers of the country on their wasteful way to the sea. This will be only an additional work for the farm engines now in operation, accomplished with but little increased expense. A single fact may illustrate the irrigating capacity of Mr. Mechi's machinery. It throws upon a field a quantity of the fertilising fluid equal to one inch of rainfall at a time, or 100 tons per imperial acre. And, as a proof of how deep it penetrates, the drains run freely with it, thus showing conclusively that the subsoil has been well saturated, a point of vital importance to the crop.

Deep tillage is another speciality that distinguished the Tiptree Farm *regime* at the beginning, in which Mr. Mechi led, and in which he has been followed by the farmers of the country, although few have come up abreast of him as yet in the system.

Here, then, are four specific departments of improvement in agricultural industry which the Alderman has introduced. Every one of them has been ridiculed as an impracticable and useless innovation in its turn. Three of them have already been adopted, and virtually incorporated with agricultural science and economy; and the fourth, or irrigation by steam power, bids fair to find as much favor, and as many adherents in the end as the others have done.

He has not only originated these improvements, or been the first to give them practical experiment, but he has laid down certain principles which will doubtless exercise much influence in shaping the industrial economy of agriculture hereafter in different countries. One of the best of these principles he puts in the form of a mathematical proposition. Thus:—As the meat is to the manure, so is the crop to the land. Tell me, he says, how much meat you make, and I will tell you how much corn you make, to the acre. Meat, then, is the starting point with him; the basis of his annual production, to which he looks for a satisfactory decision of his balance-sheet. To show the value he attaches to this element, the fact will suffice that he usually keeps 65 bullocks, cows, and calves, 100 sheep, and a number of pigs, besides his horses, making one head to every acre of his farm. With this amount of live stock he makes from £4 to £5 worth of meat per acre annually. Perhaps it would be safe to say that no other 170 acres of land in the world make more meat, manure, and grain in the year than the Tiptree Farm. In these results Mr. Mechi thinks his experiments and improvements have proved

Quod es demonstrandum.

Having gone over the farm pretty thoroughly, and noticed all the leading features of the establishment, I was requested by the foreman to enter my name in the visitor's book kept in his neat cottage parlor. It is a large volume, with the ruling running across both the wide pages; the left apportioned to name, town, country, and profession; the right to remarks of the visitor. It is truly a remarkable book of interesting

autographs and observations, which the philologist as well as agriculturist might pore over with lively satisfaction. It not only contains the names and comments of many of the most distinguished personages in Great Britain, but those of all other countries of Europe, even of Asia and Africa, as well as America. Foreign ambassadors, Continental savans, men of fame in the literary, scientific, and political world have here recorded their names and impressions in the most unique succession and blending. Here, under one date, is a party of Italian gentlemen, leaving their autographs and their observations in the softest syllables of their language. Then several German connoisseurs follow in their peculiar script, with comments worded heavily with hard-mouthed consonants. Then comes, perhaps, a single Russian nobleman, who expresses his profound satisfaction in the politest French. Next succeed three or four Spanish *Dons*, with a long fence of names attached to each, who give their views of the establishment in the grave, sonorous words of their language. Here, now, an American puts in his autograph, with his sharp, curt notion of the matter, as "first-rate." Very likely a turbaned Mufti or Singh of the Oriental world follows the New England farmer. Danish and Swedish knights prolong the procession, mingling with Australian wool-growers, Members of the French Royal Academy, Canadian timber-merchants, Dutch Mynheers, Brazilian coffee-planters, Belgian lace-makers, and the representatives of all other countries and professions in Christendom. An autograph-monger, with the mania strong upon him, of unscrupulous curiosity, armed furtively with a keen pair of scissors would be a dangerous person to admit to the presence of that big book without a policeman at his elbow.

Tiptree Hall has its own literature also, in two or three volumes, written by Mr. Mechi himself, and describing fully his agricultural experience and experiments, and giving facts and arguments which every English and American farmer might study with profit.

CHAPTER III.

ENGLISH AND AMERICAN BIRDS.
"What thou art we know not;
What is most like thee?
From rainbow clouds there flow not
Drops so bright to see,
As from thy presence showers a rain of melody."
SHELLEY'S "SKYLARK."

"Do you ne'er think what wondrous beings these?
Do you ne'er think who made them, and who taught
The dialect they speak, whose melodies
Alone are the interpreters of thought?
Whose household words are songs in many keys,
Sweeter than instrument of man e'er caught!
Whose habitations in the tree-tops, even,
Are half-way houses on the road to heaven."
LONGFELLOW.

Having spent a couple of hours very pleasantly at Tiptree Hall, I turned my face in a northerly direction for a walk through the best agricultural section of Essex. While passing through a grass field recently mown, a lark flew up from almost under my feet. And there, partially overarched by a tuft of clover, was her little all of earth—a snug, warm nest with two small eggs in it, about the size and color of those of the ground-chirping-bird of New England, which is nearer the English lark than any other American bird. I bent down to look at them with an interest an American could only feel. To him the lark is to the bird-world's companionship and music what the angels are to the spirit land. He has read and dreamed of both from his childhood up. He has believed in both poetically and pleasantly, sometimes almost positively, as real and beautiful individualities. He almost credits the poet of his own country, who speaks of hearing "the downward beat of angel wings." In his facile faith in the substance of picturesque and happy shadows, he sometimes tries to believe that the *phœnix* may have been, in some age and country, a real, living bird, of flesh and blood and genuine feathers, with long, strong wings, capable of performing the strange psychological feats ascribed to it in that most edifying picture emblazoned on the arms of Banking Companies, Insurance Offices, and Quack Doctors. He is not sure that dying swans have not sung a mournful hymn over their last moments, under an affecting and human sense of their mortality. He has believed in the English lark to the same point of pleasing credulity. Why should he not give its existence the same faith? The history of its life is as old as the English alphabet, and older still. It sang over the dark and hideous lairs of the bloody Druids centuries before Julius Cæsar was born, and they doubtless had a pleasant name for it, unless true music was hateful to their ears. It sang, without loss or change of a single note of this morning's song, to the Roman legions as they marched, or made roads in Britain. It rang the same voluntaries to the Saxons, Danes, and Normans, through the long ages, and, perhaps, tended to soften their antagonisms, and hasten their blending into one great and mighty people. How the name and song of this happiest of earthly birds run through all the rhyme and romance of English poetry, of English rural life, ever since there was an England! Take away its history and its song from her daisy-eyed meadows, and shaded lanes, and hedges breathing and blooming with sweetbrier leaves and hawthorn flowers—from her thatched cottages, veiled with ivy—from the morning tread of the reapers, and the mower's lunch of bread and cheese under the meadow elm, and you take away a living and beautiful spirit more charming than music. You take away from English poetry one of its pleiades, and bereave it of a companionship more intimate than that of the nearest neighborhood of the stars above. How the lark's life and song blend, in the rhyme of the poet, with "the sheen of silver fountains leaping to the sea," with morning sunbeams and noontide thoughts, with the sweetest breathing flowers, and softest breezes, and busiest bees, and greenest leaves, and happiest human industries, loves, hopes, and aspirations!

The American has read and heard of all this from his youth up to the day of setting his foot, for the first time, on English ground. He has tried to believe it, as in things seen, temporal and tangible. But in doing this he has to contend with a sense or suspicion of unreality—a feeling that there has been great poetical exaggeration in the matter. A patent fact lies at the bottom of this incredulity. The forefathers of New England carried no wild bird with them to sing about their cabin homes in the New World. But they found beautiful and happy birds on that wild continent, as well-dressed, as graceful in form and motion, and of as fine taste for music and other accomplishments, as if they and their ancestors had sung before the courts of Europe for twenty generations. These sang their sweet songs of welcome to the Pilgrims as they landed from the "Mayflower." These sang to them cheerily, through the first years and the later years of their stern trials and tribulations. These built their nests where the blue eyes of the first white children born in the land could peer in upon the speckled eggs with wonder and delight. What wonder that those strong-hearted puritan fathers and mothers, who

"Made the aisles of the dim wood ring
With the anthems of the free,"

should love the fellowship of these native singers of the field and forest, and give them names their hearts loved in the old home land beyond the sea! They did not consult Linnæus, nor any musty Latin genealogy of Old World birds, at the christening of these songsters. There was a good family resemblance in many cases. The blustering partridge, brooding over her young in the thicket, was very nearly like the same bird in England. For the mellowthroated thrush of the old land they found a mate in the new, of the same size, color, and general habits, though less musical. The blackbird was nearly

the same in many respects, though the smaller American wore a pair of red epaulettes. The swallows had their coat tails cut after the same old English pattern, and built their nests after the same model, and twittered under the eaves with the same ecstasy, and played the same antics in the air. But the two dearest home-birds of the fatherland had no family relations nor counterparts in America; and the pilgrim fathers and their children could not make their humble homes happy without the lark and the robin, at least in name and association; so they looked about them for substitutes. There was a plump, full-chested bird, in a chocolate-colored vest, with a bluish dress coat, that would mount the highest tree-top in early spring, and play his flute by the hour for very joy to see the snow melt and the buds swell again. There was such a rollicking happiness in his loud, clear notes, and he apparently sang them in such sympathy with human fellowships, and hopes, and homes, and he was such a cheery and confiding denizen of the orchard and garden withal, that he became at once the pet bird of old and young, and was called the *robin*; and well would it be if its English namesake possessed its sterling virtues; for, with all its pleasant traits and world-wide reputation, the English robin is a pretentious, arrogant busybody, characteristically pugilistic and troublesome in the winged society of England. In form, dress, deportment, disposition, and in voice and taste for vocal music, the American robin surpasses the English most decidedly. In this our grave forefathers did more than justice to the homebird they missed on Plymouth Rock. In this generous treatment of their affection for it, they perhaps condoned for mating the English lark so incongruously; but it was true their choice was very limited. To match the *prima donna carissima* of English field and sky, it was necessary to select a meadow bird, with some other features of resemblance. It would never do to give the cherished name and association to one that lived in the forest, or built its nest in the tree-tops or house-tops, or to one that was black, yellow, or red. Having to conciliate all these conditions, and do the best with the material at hand, they pitched upon a rather large, brownish bird, in a drab waistcoat, slightly mottled, and with a loud, cracked voice, which nobody ever liked. So it never became a favorite, even to those who first gave it the name of lark. It was not its only defect that it lacked an ear and voice for music. There is always a scolding accent that marks its conversation with other birds in the brightest mornings of June. He is very noisy, but never merry nor musical. Indeed, compared with the notes of the English lark, his are like the vehement ejaculations of a maternal duck in distress.

Take it in all, no bird in either hemisphere equals the English lark in heart or voice, for both unite to make it the sweetest, happiest, the welcomest singer that was ever winged, like the high angels of God's love. It is the living ecstacy of joy when it mounts up into its "glorious privacy of light." On the earth it is timid, silent, and bashful, as if not at home, and not sure of its right to be there at all. It is rather homely withal, having nothing in feather, feature, or form, to attract notice. It is seemingly made to be heard, not seen, reversing the old axiom addressed to children when getting voicy. Its mission is music, and it floods a thousand acres of the blue sky with it several times a day. Out of that palpitating speck of living joy there wells forth a sea of twittering ecstacy upon the morning and evening air. It does not ascend by gyrations, like the eagle or birds of prey. It mounts up like a human aspiration. It seems to spread out its wings and to be lifted straight upwards out of sight by the afflatus of its own happy heart. To pour out this in undulating rivulets of rhapsody is apparently the only motive of its ascension. This it is that has made it so loved of all generations. It is the singing angel of man's nearest heaven, whose vital breath is music. Its sweet warbling is only the metrical palpitation of its life of joy. It goes up over the roof-trees of the rural hamlet on the wings of its song, as if to train the rural soul to trial flights heavenward. Never did the Creator put a voice of such volume into so small a living thing. It is a marvel—almost a miracle. In a still hour you can hear it at nearly a mile's distance. When its form is lost in the hazy lace-work of the sun's rays above, it pours down upon you all the thrilling semitones of its song as distinctly as if it were warbling to you in your window.

The only American bird that could star it with the English lark, and win any admiration at a popular concert by its side, is our favourite comic singer, the *Bobolink*. I have thought often, when listening to British birds at their morning rehearsals, what a sensation would ensue if Master Bob, in his odd-fashioned bib and tucker, should swagger into their midst, singing one of those Low-Dutch voluntaries which he loves to pour down into the ears of our mowers in haying time. Not only would such an apparition and overture throw the best-trained orchestra of Old World birds into amazement or confusion, but astonish all the human listeners at an English concert. With what a wonderment would one of these blooming, country milkmaids look at the droll harlequin, and listen to those familiar words of his, set to his own music:-

Go to milk! go to milk!
Oh, Miss Phillisey,
Dear Miss Phillisey,
What will Willie say
If you don't go to milk!
No cheese, no cheese,
No butter nor cheese
If you don't go to milk.

It is a wonder that in these days of refined civilization, when Jenny Lind, Grisi, Patti, and other celebrated European singers, some of them from very warm climates, are transported to America to delight our Upper-Tendom, that there should be no persistent and successful effort to introduce the English lark into our out-door orchestra of singing-birds. No European voice would be more welcome to the American million. It would be a great gain to the nation, and be helpful to our religious devotions, as well as to our secular satisfactions. In several of our Sabbath hymns there is poetical reference to the lark and its song. For instance, that favorite psalm of gratitude for returning Spring opens with these lines:—

"The winter is over and gone,
The thrush whistles sweet on the spray,
The turtle breathes forth her soft moan,
The *lark* mounts on high and warbles away."

Now, not one American man, woman, or child in a thousand ever heard or saw an English lark, and how is he, she, or it to sing the last line of the foregoing verse with the spirit and understanding due to an exercise of devotion? The American lark never mounts higher than the top of a meadow elm, on which it see-saws, and screams, or quacks, till it is tired; then draws a bee-line for another tree, or a fence-post, never even undulating on the voyage. It may be said, truly enough, that the hymn was written in England. Still, if sung in America from generation to generation, we ought to have the English lark with us, for our children to see and hear, lest they may be tempted to believe that other and more serious similes in our Sabbath hymns are founded on fancy instead of fact.

Nor would it be straining the point, nor be dealing in poetical fancies, if we should predicate upon the introduction of the English lark into American society a supplementary influence much needed to unify and nationalise the heterogeneous elements of our population. Men, women, and children, speaking all the languages and representing all the countries and races of Europe, are streaming in upon us weekly in widening currents. The rapidity with which they become assimilated to the native population is remarkable. But there is one element from abroad that does not Americanise itself so easily—and that, curiously, is one the most American that comes from Europe—in other words, the *English*. They find with us everything as English as it can possibly be out of England—their language, their laws, their literature, their very bibles, psalmbooks, psalm-tunes, the same faith and forms of worship, the same common histories, memories, affinities, affections, and general structure of social life and public institutions; yet they are generally the very last to be and feel at home in America. A Norwegian mountaineer, in his deerskin doublet, and with a dozen English words picked up on the voyage, will *Americanise* himself more in one year on an Illinois prairie than an intelligent, middle-class Englishman will do in ten, in the best society of Massachusetts. Now, I am not dallying with a facetious fantasy when I express the opinion, that the life and song of the English lark in America, superadded to the other institutions and influences indicated, would go a great way in fusing this hitherto insoluble element, and blending it harmoniously with the best vitalities of the nation. And this consummation would well repay a special and extraordinary effect. Perhaps this expedient would be the most successful of all that remain untried. A single incident will prove that it is more than a mere theory. Here it is, in substance:—

Some years ago, when the Australian gold fever was hot in the veins of thousands, and fleets of ships were conveying them to that far-off, uncultivated world, a poor old woman landed with the great multitude of rough and reckless men, who were fired to almost frenzy by dreams of ponderous nuggets and golden fortunes. For these they left behind them all the enjoyments, endearments, all the softening sanctities and surroundings of home and social life in England. For these they left mothers, wives, sisters and daughters. There they were, thinly tented in the rain, and the dew, and the mist, a busy, boisterous, womanless camp of diggers and grubbers, roughing-and-tumbling it in the scramble for gold mites, with no quiet Sabbath breaks, nor Sabbath songs, nor Sabbath bells to measure off and sweeten a season of rest. Well, the poor widow, who had her cabin within a few miles of "the diggings," brought with her but few comforts from the old homeland—a few simple articles of furniture, the bible and psalm-book of her youth, and an English lark to sing to her solitude the songs that had cheered

her on the other side of the globe. And the little thing did it with all the fervor of its first notes in the English sky. In her cottage window it sang to her hour by hour at her labor, with a voice never heard before on that wild continent. The strange birds of the land came circling around in their gorgeous plumage to hear it. Even four-footed animals, of grim countenance, paused to hear it. Then, one by one, came other listeners. They came reverently, and their voices softened into silence as they listened. Hard-visaged men, bare-breasted and unshaven, came and stood gentle as girls; and tears came out upon many a tanned and sun-blistered cheek as the little bird warbled forth the silvery treble of its song about the green hedges, the meadow streams, the cottage homes, and all the sunny memories of the fatherland. And they came near unto the lone widow with pebbles of gold in their hard and horny hands, and asked her to sell them the bird, that it might sing to them while they were bending to the pick and the spade. She was poor, and the gold was heavy; yet she could not sell the warbling joy of her life. But she told them that they might come whenever they would to hear it sing. So, on Sabbath days, having no other preacher nor teacher, nor sanctuary privilege, they came down in large companies from their gold-pits, and listened to the devotional hymns of the lark, and became better and happier men for its music.

Seriously, it may be urged that the refined tastes, arts, and genius of the present day do not develop themselves symmetrically or simultaneously in this matter. Here are connoisseurs and enthusiasts in vegetable nature hunting up and down all the earth's continents for rare trees, plants, shrubs, and flowers. They are bringing them to England and America in shiploads, to such extent and variety, that nearly all the dead languages and many of the living are ransacked to furnish names for them. Llamas, dromedaries, Cashmere goats, and other strange animals, are brought, thousands of miles by sea and land, to be acclimatised and domesticated to

these northern countries. Artificial lakes are made for the cultivation of fish caught in Antipodean streams. That is all pleasant and hopeful and proper. The more of that sort of thing the better. But why not do the other thing, too? Vattemare made it the mission of his life to induce people of different countries to exchange books, or unneeded duplicates of literature. We need an Audubon or Wilson, not to make new collections of feathered skeletons, and new volumes on ornithology, but to effect an exchange of living birds between Europe and America; not for caging, not for Zoological gardens and museums, but for singing their free songs in our fields and forests. There is no doubt that the English lark would thrive and sing as well in America as in this country. And our bobolink would be as easily acclimatised in Europe. Who could estimate the pleasure which such an exchange in the bird-world would give to millions on both sides of the Atlantic?

There are some English birds which we could not introduce into the feathered society of America, any more than we could import a score of British Dukes and Duchesses, with all their hereditary dignities and grand surroundings, into the very heart and centre of our democracy. For instance, the grave and aristocratic rooks, if transported to our country, would turn up their noses and caw with contempt at our institutions—even at our oldest buildings and most solemn and dignified oaks. It is very doubtful if they would be conciliated into any respect for the Capitol or The White House at Washington. They have an intuitive and most discriminating perception of antiquity, and their adhesion to it is invincible. Whether they came in with the Normans, or before, history does not say. One thing would seem evident. They are older than the Order of the Garter, and belonged to feudalism. They are the living spirits of feudalism, which have survived its human retainers by several hundred years, and now represent the defunct institution as pretentiously as in King Stephen's day. They are as fond of old Norman castles, cathedrals, and churches, as the very ivy itself, and cling to them with as much pertinacity. For several hundred generations of bird-life, they and their ancestors have colonised their sable communities in the baronial park-trees of England, and their descendants promise to abide for as many generations to come. In size, form, and color they differ but little from the American crow, but are swifter on the wing, with greater "gift of the gab," and less dignified in general deportment, though more given to aristocratic airs. Although they emigrated from France long before "*La Democratic Sociale*" was ever heard of in that country, they may be considered the founders of the *Socialistic* theory and practice; and to this day they live and move in *phalansteries*, which succeed far better than those attempted by the American "*Fourierites*" some years ago. As in human communities, the collision of mind with mind contributes fortuitous scintillations of intelligence to their general enlightenment; so gregarious animals, birds and bees seem to acquire especial quick-wittedness from similar intercourse. The English rook, therefore, is more astute, subtle, and cunning than our American crow, and some of his feats of legerdemain are quite vulpine.

The jackdaw is to the rook what the Esquimaux is to the Algonquin Indian; of the same form, color, and general habits, but smaller in size. They are as fond of ancient abbeys and churches as ever were the monks of old. Indeed, they have many monkish habits and predilections, and chatter over their Latin rituals in the storied towers of old Norman cathedrals, and in the belfries of ivy-webbed churches in as vivicacious confusion.

There is no country in the world of the same size that has so many birds in it as England; and there are none so musical and merry. They all sing here congregationalwise, just as the people do in the churches and chapels of all religious denominations. As these buildings were fashioned in early times after the Gothic order of elm and oak-tree architecture, so the human worshippers therein imitated the birds, as well as the branches,

of those trees, and learned to sing their Sabbath hymns together, young and old, rich and poor, in the same general uprising and blending of multitudinous voices. I believe everything sings that has wings in England. And well it might, for here it is safe from shot, stones, snares, and other destructives. "Young England" is not allowed to sport with firearms, after the fashion of our American boys. You hear no juvenile popping at the small birds of the meadow, thicket, or hedge-row, in Spring, Summer, or Autumn. After travelling and sojourning nearly ten years in the country, I have never seen a boy throw a stone at a sparrow, or climb a tree for a bird's-nest. The only birds that are not expected to die a natural death are the pheasant, partridge, grouse, and woodcock; and these are to be killed according to the strictest laws and customs, at a certain season of the year, and then only by titled or wealthy men who hold their vested interest in the sport among the most rigid and sacred rights of property. Thus law, custom, public sentiment, climate, soil, and production, all combine to give bird-life a development in England that it attains in no other country. In no other land is it so multitudinous and musical; in none is there such ample and varied provision for housing and homing it. Every field is a great bird's-nest. The thick, green hedge that surrounds it, and the hedge-trees arising at one or two rods' interval, afford nesting and refuge for myriads of these meadow singers. The groves and thickets are full of them and their music; so full, indeed, that sometimes every leaf seems to pulsate with a little piping voice in the general concert. Nor are they confined to the fields, groves, and hedges of the quiet country. If the census of the sparrows alone in London could be taken, they would count up to a larger figure than all the birds of a New England county would reach. Then there is another interesting feature of this companionship. A great deal of it lasts through the entire year. There are ten times as many birds in England as in America in the winter. Here the fields are green through the coldest months. No deep and drift-

ing snows cover a frozen earth for ten or twelve weeks, as with us. There is plenty of shelter and seeds for birds that can stand an occasional frost or wintry storm, and a great number of them remain the whole year around the English homesteads.

If such a difference were a full compensation, our North American birds make up in dress what they fall short of English birds in voice and musical talent. The robin redbreast and the goldfinch come out in brighter colors than any other beaux and belles of the season here; but the latter is only a slender-waisted brunette, and the former a plump, strutting, little coxcomb, in a mahogany-colored waistcoat. There is nothing here approaching in vivid colors the New England yellow-bird, hang-bird, red-bird, indigo-bird, or even the bluebird. In this, as well as other differences, Nature adjusts the system of compensation which is designed to equalise the conditions of different countries.

CHAPTER IV.

TALK WITH AN OLD MAN ON THE WAY—OLD HOUSES IN ENGLAND—THEIR AMERICAN RELATIONSHIPS—ENGLISH HEDGES AND HEDGE-ROW TREES—THEIR PROBABLE FATE—CHANGE OF RURAL SCENERY WITHOUT THEM.

From Tiptree I had a pleasant walk to Coggeshall, a unique and antique town, marked by the quaint and picturesque architecture of the Elizabethan *regime*. On the way I met an old man, eighty-three years of age, busily at work with his wheel-barrow, shovel, and bush-broom, gathering up the droppings of manure on the road. I stopped and had a long talk with him, and learned much of those ingenious and minute industries by which thousands of poor men house, feed, and clothe themselves and their families in a country super-abounding with labor. He had nearly filled his barrow, after trundling it for four miles.

He could sell his little load for 4d. to a neighboring farmer; but he intended to keep it for a small garden patch allotted to him by his son, with whom he lived. These few square yards of land constituted the microscopic point of his attachment to that great globe still holding in reserve unmeasured territories of productive soil, on which nor plough, nor spade, nor human foot, nor life has ever left a lasting mark. These made his little farm, as large to him and to his octogenarian sinews and ambitions as was the Tiptree Estate to Alderman Mechi. It filled his mind with as busy occupation and as healthy a stimulus. That rude barrow, with its clumsy wheel, thinly rimmed with an iron hoop, was to him what the steam engine, and two miles of iron tubing, and all its hose-power were to that eminent agriculturist, of whom he spoke in terms of high esteem as a neighbor, and even as a competitor. Proportionately they were on the same footing; the one with his 170 square acres, the other with his 170 square feet. It was pleasant and instructive to hear him speak with such sunny and cheery hope of his earthly lot and doings. His son was kind and good to him. He could read, and get many good books. He ate and slept well. He was poor but comfortable. He went to church on Sunday, and thought much of heaven on week days. His cabbages were a wonder; some with heads as large as a half-bushel measure. He did something very respectable in the potato and turnip line. He had grown beans and beets which would show well in any market. He always left a strip or corner for flowers. He loved to grow them; they did him good, and stirred up young-man feelings in him. He went on in this way with increased animation, following the lead of a few questions I put in occasionally to give direction to the narrative of his experience. How much I wished I could have photographed him as he stood leaning on his shovel, his wrinkled face and gray, thin hair, moistened with perspiration, while his coat lay inside out on one of the handles of his barrow! The July sun, that warmed him at his work, would have made an interest-

ing picture of him, if some one could have held a camera to its eye at the moment. I added a few pennies to his stock-in-trade, and continued my walk, thinking much of that wonderful arrangement of Providence by which the infinite alternations and gradations of human life and condition are adjusted; fitting a separate being, experience, and attachment to every individual heart; training its tendrils to cling all its life long to one slightly individualised locality, which another could never call home; giving itself and all its earthly hopes to an occupation which another would esteem a prison discipline; sucking the honey of contentment out of a condition which would be wormwood to another person on the same social level.

On reaching Coggeshall, I became again the guest of a Friend, who gave me the same old welcome and hospitality which I have so often received from the members of that society. After tea, he took me about the town, and showed me those buildings so interesting to an American—low, one-story houses, with thatched roofs, clay-colored, wavy walls, rudely-carved lintels, and iron-sash windows opening outward on hinges like doors, with squares of glass 3 inches by 4;—houses which were built before the keel of the Mayflower was laid, which conveyed the Pilgrims to Plymouth Rock. Here, now! see that one on the other side of the street, looking out upon a modern and strange generation through two ivy-browed eyes just lighted up to visible speculation by a single candle on the mantel-piece! A very animated and respectable baby was carried out of that door in its mother's arms, and baptised in the parish church, before William Shakespeare was weaned. There is a younger house near by, which was a century old when Washington was born. These unique, old dwellings of town, village, and hamlet in England, must ever possess an interest to the American traveller which the grand and majestic cathedrals, that fill him with so much admiration, cannot inspire. We link the life of our nation more directly to these humbler

buildings. Our forefathers went out of these houses to the New World. The log huts they first erected served them and their families as homes for a few years; then were given to their horses and cattle for stabling; then were swept away, as too poor for either man or beast. The second generation of houses made greater pretensions to comfort, and had their day, then passed away. They were nearly all one-story, wooden buildings, with a small apartment on each side of a great chimney, and a little bed-roomage in the garret for children. Then followed the large, red, New England mansion, broadside to the road, two stories high in front, with nearly a rood of back roof declining to within five or six feet of the ground, and covering a great, dark kitchen, flanked on one side by a bed-room, and on the other by the buttery. A ponderous chimney arose out of the middle of the building, giving a fire-place of eight feet back to the kitchen, and one of half the dimensions to each of the other two large rooms—the *north* and *south*. For, like the republic they founded, its forefathers and ours divided their dwellings by a kind of Mason and Dixon's Line, into two parts, giving them these sectional appellations which have represented such antagonisms and made us such trouble. Every one of these old-fashioned houses had its "North" and "South" rooms on the ground-floor, and duplicates, of the same size and name, above, divided by the massive, hollow tower, called a chimney. A double front door, with panels, scrolled with rude carving, opened right and left into the portly building, which, in the *tout ensemble*, looked like a New England gentleman of the olden time, in his cocked hat, and hair done up in a *queue*. These were the houses built "when George the Third was King. " In these were born the men of the American Revolution. They are the oldest left in the land; and, like the Revolutionary pensioners, they are fast disappearing. In a few years, it will be said the last of them has been levelled to the ground, just as the paragraph will circulate through the newspapers that the last soldier of the War of Independence is dead.

Thus, the young generation in America, now reciting in our schools the rudimental facts of the common history of the English-speaking race, will come to the meridian of manhood at a time when the three first generations of American houses shall have been swept away. But, travelling over a space of three centuries' breadth, they will see, in these old English dwellings, where the New World broke off from the Old—the houses in which the first settlers of New England were born; the churches and chapels in which they were baptised, and the school-houses in which they learned the alphabet of the great language that is to fill the earth with the speech of man's rights and God's glory. One hundred millions, speaking the tongue of Shakespeare and Milton on the American continent, and as many millions more on continents more recently settled by the same race, across the ocean, and across century-seas of time, shall moor their memories to these humble dwellings of England's hamlets, and feel how many taut and twisted liens attach them to the motherland of mighty nations.

On reckoning up the log of my first day's walk, I found I had made full twelve miles by road and field; and was more than satisfied with such a trial of country air and exercise, and with the enjoyment of its scenery and occupations. The next day I made a longer distance still, from Coggeshall to Great Bardfield, or about eighteen miles; and felt at the end that I had established a reasonable claim to convalescence. The country on the way was marked by the quiet and happy features of diversified plenty. The green and gold of pastures, meadows, and wheat-fields; the picturesque interspersion of cottages, gardens, stately mansions, parks and lawns, all enlivened by a well-proportioned number of mottled cows feeding or lying along the brook-banks, and sheep grazing on the uplands,—all these elements of rural life and scenery were blended with that fortuitous felicity which makes the charm of Nature's country pictures.

At Bardfield I was again homed for the night by a Friend; and after tea made an evening walk with him about the farm of a member of the same society, living in the outskirts of the town, who cultivates about 400 acres of excellent land, and is considered one of the most practical and successful agriculturists of Essex. His fields were larger and fewer than I had noticed on my walk in a farm of equal size. This feature indicates the modern improvements in English farming more prominently to the cursory observer than any other that attracts his eye. It is a rigidly utilitarian innovation on the old system, that does not at all promise to improve the picturesque aspect of the country. To "reconstruct the map" of a county, by wire-fencing it into squares of 100 acres each, after grubbing up all the hedges and hedge-trees, would doubtless add seven and a quarter per cent. to the agricultural production of the shire, and gratify many a Gradgrind of materialistic economy; but who would know England after such a transformation? One would be prone to reiterate Patrick's exclamation of surprise, when he first shouldered a gun and tested the freedom of the forest in America. Seeing a small bird in the top of a tree, he pointed the fowling-piece in that direction, turned away his face, and fired. A tree-toad fell to the ground from an agitated branch. The exulting Irishman ran and picked it up in triumph, and held it out at arm's length by one of its hind legs, exclaiming, "And how it alters a bird to shoot its feathers off, to be sure!" It would alter England nearly as much in aspect, if the unsparing despotism of £ s. d. should root out the hedge-row trees, and substitute invisible lines of wire for the flowering hawthorn as a fencing for those fields which now look so much like framed portraits of Nature's best painting.

The tendency of these utilitarian times may well occasion an unpleasant concern in the lovers of English rural scenery. What changes may come in the wake of the farmer's steam-engine, steam-plough, or under the smoke-shadows from his factory-like chimney, these recent "improvements" may sug-

gest and induce. One can see in any direction he may travel these changes going on silently. Those little, unique fields, defined by lines and shapes unknown to geometry, are going out of the rural landscape. And when they are gone, they will be missed more than the amateurs of agricultural artistry imagine at the present moment. What some one has said of the peasantry, may be said, with almost equal deprecation, of these picturesque tit-bits of land, which,—

"Once destroyed, never can be supplied."

And destroyed they will be, as sure as science. As large farms are swallowing up the little ones between them, so large fields are swallowing these interesting patches, the broad-bottomed hedging of which sometimes measures as many square yards as the space it encloses.

There is much reason to fear that the hedge-trees will, in the end, meet with a worse fate still. Practical farmers are beginning to look upon them with an evil eye—an eye sharp and severe with pecuniary speculation; that looks at an oak or elm with no artist's reverence; that darts a hard, dry, timber-estimating glance at the trunk and branches; that looks at the circumference of its cold shadow on the earth beneath, not at the grand contour and glorious leafage of its boughs above. The farmer who was taking us over his large and highly-cultivated fields, was a man of wide intelligence, of excellent tastes, and the means wherewithal to give them free scope and play. His library would have satisfied the ambition of a student of history or belles-lettres. His gardens, lawn, shrubbery, and flowers would grace the mansion of an independent gentleman. He had an eye to the picturesque as well as practical. But I could not but notice, as significant of the tendency to which I have referred, that, on passing a large, outbranching oak standing in the boundary of two fields, he remarked that the detriment of its shadow could not have been less than ten shillings a year for half a century. As we proceeded from field to field, he recurred to the same subject by calling our attention to the circumference of the shadow cast on the best land of the farm by a thrifty, luxuriant ash, not more than a foot in diameter at the butt. Up to the broad rim of its shade, the wheat on each side of the hedge was thick, heavyheaded and tall, but within the cool and sunless circle the grain and grass were so pale and sickly that the bare earth would have been relief to a farmer's eye.

The three great, distinctive graces of an English landscape are the hawthorn hedges, the hedge-row trees, and the everlasting and unapproachable greenness of the grass-fields they surround and embellish. In these beautiful features, England surpasses all other countries in the world. These make the peculiar charm of her rural scenery to a traveller from abroad. These are the salient lineaments of Motherland's face which the memories of myriads she has sent to people countries beyond the sea cling to with such fondness; memories that are transmitted from generation to generation; which no political revolutions nor severances affect; which are handed down in the unwritten legends of family life in the New World, as well as in the warp and woof of American literature and history. Will the utilitarian and unsparing science of these latter days, or of the days to come, shear away these beautiful tresses, and leave the brow and temples of the Old Country they have graced bare and brown under the bald and burning sun of material economy? It is not an idle question, nor too early to ask it. It is a question which will interest more millions of the English race on the American continent than these home-islands will ever contain. There are influences at work which tend to this unhappy issue. Some of these have been already indicated, and others more powerful still may be mentioned.

Agriculture in England has to run the gauntlet of many pressing competitions, and carry a heavy burden of taxation as it runs. These will be noticed, hereafter, in their proper connection. Farming, therefore, is being reduced to a rigid science. Every acre of land must be put up to its last ounce of production. Every square foot of it must be utilised to the growth of something for man and beast.

Manures for different soils are tested with as much chemical precision as ever was quinine for human constitutions. Dynameters are applied to prove the power of working machinery. Labor is scrutinised and economised, and measured closely up to the value of a farthing's-worth of capacity. A shilling's difference per acre in the cost of ploughing by horse-flesh or steam brings the latter into the field. The sound of the flail is dying out of the land, and soon will be heard no more. Even threshing machines worked by horses are being discarded, as too slow and old-fashioned. Locomotive steam-engines, on broad-rimmed wheels, may be met on the turnpike road, travelling on their own legs from farm to farm to thresh out wheat, barley, oats, and beans, for a few pence per bushel. They make nothing of ascending a hill without help, or of walking across a ploughed field to a rick-yard. Iron post and rail fencing, in lengths of twenty feet on wheels, drawn about by a donkey, bids fair to supersede the old wooden hurdles for sheep fed on turnips or clover. It is an iron age, and wire fencing is creeping into use, especially in the most scientifically cultivated districts of Scotland, where the elements and issues of the farmer's balance-sheet are looked to with the most eager concern. Iron wire grows faster than hawthorn or buckthorn. It doubtless costs less. It needs no yearly trimming, like shrubs with sap and leaves. It does not occupy a furrow's width as a boundary between two fields. It may be easily transposed to vary enclosures. It is not a nesting place for destructive birds or vermin. These and other arguments, of the same utilitarian genus, are making perceptible headway. Will they ever carry the day against the green hedges? I think they would, very soon, if the English farmer owned the land he cultivates. But such is rarely the case. Still, this fact may not prevent the final consummation of this policy of material interest. In a great many instances, the tenant might compromise with the landlord in such a way as to bring about this "modern improvement." And a compar-

atively few instances, showing a certain per centage of increased production per acre to the former, and a little additional rentage to the latter, would suffice to give the innovation an impulse that would sweep away half the hedges of the country, and deface that picture which so many generations have loved to such enthusiasm of admiration.

Will the trees of the hedge-row be exposed to the same end? I think they will. Though trees are the most sacred things the earth begets in England, as has already been said, the farmer here looks at them with an evil eye, as horse-leeches that bleed to death long stretches of the land he pays £2 per acre for annually to his landlord. The hedge, however wide-bottomed, is his fence; and fencing he must have. But these trees, arising at narrow intervals from the hedge, and spreading out their deadening shades upon his wheatfields on either side, are not useful nor ornamental to him. They may look prettily, and make a nice picture in the eyes of the sentimental tourist or traveller, but he grudges the ground they cover. He could well afford to pay the landlord an additional rentage per annum more than equal to the money value of the yearly growth of these trees. Besides, the landlord has, in all probability, a large park of trees around his mansion, and perhaps compact plantations on land unsuited to agriculture. Thus the high value of these hedge-row trees around the fields of his tenant, which he will realise on the spot, together with some additional pounds in rent annually to himself and heirs, would probably facilitate this levelling arrangement in face of all the restrictions that the law of entail might seem to throw in the way.

If, therefore, the hedges of England disappear before the noiseless and furtive progress of utilitarian science, the trees that rise above them in such picturesque ranks will be almost certain to go with them. Then, indeed, a change will come over the face of the country, which will make it difficult for one to recognise it who daguerreotyped its most beautiful features upon his memory before they were obliterated by these latter-day "improvements."

CHAPTER V.

*A FOOTPATH WALK AND ITS INCI-
DENTS—HARVEST ASPECTS—ENG-
LISH AND AMERICAN SKIES—HUM-
BLER OBJECTS OF CONTEMPLA-
TION—THE DONKEY: ITS USES AND
ABUSES.*

Immediately after breakfast the following morning, my kind host accompanied me for a mile on my walk, and
put me on a footpath across the fields,
by which I might save a considerable
distance on the way to Saffron Walden,
where I proposed to spend the Sabbath.
After giving me minute directions as to
the course I was to follow, he bade me
good-bye, and I proceeded on at a brisk
pace through fields of wheat and clover,
greatly enjoying the scenery, the air,
and exercise. Soon I came to a large
field quite recently ploughed up *clean*,
footpath and all. Seeing a gate at each
of the opposite corners, I made my way
across the furrows to the one at the left,
as it seemed to be more in the direction
indicated by my host. There the path
was again broad and well-trodden, and I
followed it through many fields of grain
yellowing to the harvest, until it opened
into the main road. This bore a little
more to the left than I expected, but, as I
had never travelled it before, I believed
it was all right. Thaxted was half way
to Saffron Walden, and there I had intended to stop an hour or two for dinner and rest, then push on to the end
of the day's walk as speedily as possible. At about noon, I came suddenly
down upon the town, which seemed remarkably similar to the one I had left, in
size, situation, and general features. The
parish church, also, bore a strong resemblance to the one I had noticed the previous evening. These old Essex towns
are "as much alike as two peas," and
you must make a note of it, as Captain
Cuttle says, was the thought first suggested by the coincidence. I went into a cosy, clean-faced inn on the main
street, and addressed myself with much
satisfaction to a short season of rest and
refreshment, exchanging hot and dusty
boots for slippers, and going through
other preliminaries to a comfortable
time of it. Rang the bell for dinner, but
before ordering it, asked the waiting-
maid, with a complacent idea that I had
improved my walking pace, and made
more than half the way—

"How far is it to Saffron Walden?"

"Twelve miles, sir."

"Twelve miles, indeed! Why, it is only twelve miles from Great Bardfield!"

"Well, this is Great Bardfield, sir."

"Great Bardfield! What! How is this?
What do you mean?"

She meant what she said, and it was
as true as two and two make four; and
she was not to be beaten out of it by
a stare of astonishment, however a dis-
comfited man might expand his eyes
with wonder, or cloud his face with cha-
grin. It was a patent fact. There, on the
opposite side of the street, was the
house in which I slept the night before;
and here, just coming up to the door of
the inn, was the good lady of my host.
Her form and voice, and other identi-
fications dispelled the mist of the mistake; and it came out as clear as day that
I had followed the direction of my host,
to bear to the left, far too liberally, and
that I had been walking at my best speed
in a "vicious circle" for full two hours
and a half, and had landed just where I
commenced, at least within the breadth
of a narrow street of the same point.

My good friends urged me to stop
and dine with them, and then make a
fair start for the end of my week's journey. But it was still twelve miles to Saffron Walden, and I was determined to
put half of them behind me before dinner. So, taking a second leave of them in
the course of three hours, I set out again
on my walk, a wiser man in the practical understanding of the proverb, "The
longest way around is the shortest way
there." At 2 p.m. I reached Thaxted, and
rectified my first notion of the town,
formed when I mistook it for Bardfield.
Having made six miles extra between
the two points, I resumed my walk after
a short delay at the latter.

The weather was glorious. A cloud-
less sun shone upon a little sky-crys-
talled world of beauty, smaller in every
dimension than you ever see in America. And this is a feature of English
scenery that will strike the American
traveller most impressively at the first
glance, whether he looks at it by night
or day. It is not that Nature, in adjusting
the symmetries of her scenic structures,
nicely apportions the skyscape to the
landscape of a country merely for artistic effect. It is not because the island of
Great Britain is so small in circumference that the sky is proportioned to it,
as the crystal is to the dial of a watch;
that it is so apparently low; that the stars
it holds to its moist, blue bosom are so
near at midnight, and the sun so large
at noon. It comes, doubtless, from that
constant humidity of the atmosphere
which distinguishes the climate of England, and gives to both land and sky an
aspect which is quite unknown to our
great western continent. An American,
after having habituated himself to this
aspect, on returning to his own country,
will be almost surprised at a feature of
its scenery which he never noticed before. He will be struck at the loftiness
of the sky; at the vividness of its blue
and gold, the sharp, unsoftened light of
the stars, and, as it were, the contracted
pupil of the sun's eye at mid-day. The
sunset glories of our western heavens
play upon a ground of rigid blue. "The
Northern Lights," which, at their winter
evening illuminations, seem to have
shredded into wavy filaments all the
rainbows that have spanned the cham-
bers of the East since the Flood, and
to upspring, in mirthful fantasy, to hang
their infinitely-tinted tresses to the
zenith's golden diadem of stars—even
they sport upon the same lofty concave
of dewless blue, which looks through
and through the lacework and everchanging drapery of their mingled hues
in the most witching mazes of their
nightly waltz, giving to each a definiteness that our homely Saxon tongue
might fit with a name.

But here, on the lower grounds of instructive meditation, is a humbler individuality of the country to notice. Here

is the most sadly abused and melancholy living creature in all England's animal realm that meets me in the midst of these reflections on things supernal and glorious. I will let the Northern Lights go, with their gorgeous pantomimes and midnight revelries, and have a moment's communing with this unfortunate quadruped. It is called in derision here a "*donkey*," but an ass, in a more generous time, when one of his race and size bore upon his back into the Holy City the world's Saviour and Re-Creator. Poor, libelled, hopeless beast! I pity you from my heart's heart. How I wish for Sterne's pen to do you some measure of justice or condolence under this heavy load of opprobrium that bends your back and makes your life so sunless and bitter! Come here, sir!— here is a biscuit for you, of the finest wheat; few of your race get such morsels; so, eat it and be thankful. What ears! No wonder our friend Patrick called you "the father of all rabbits" at first sight. No! don't turn away your head, as if I were going to strike you.

Most animals are best described from a certain point of *view*,—in a fixed and quiescent attitude. But the donkey should be taken in the very act of this characteristic motion. You put out your hand in the gentlest manner to pat any one of them you meet, and he will instinctively turn away his head for fear of a beating.

There is an interesting speculation now coming up among modern reveries in regard to the immortality of certain animals of great intelligence and domestic virtues. A large and tender kindness of disposition is the father of the thought, it may be; but the thought seems to gain ground and take shape, that so much of apparently human mind and heart as the dog possesses cannot be destined to annihilation at his death, but must live and enlarge in another sphere of existence. Having thus opened, if it may be said reverently, a back-door into immortality for sagacious and affectionate dogs and horses, they leave it ajar for the admission of animals of less intelligence—even for all the kinds that Noah took into the ark, perhaps, al-

though the theory is still nebulous and undefined. Now, I would beg the kind-hearted adherents to this theory not to think I am seeking to play off a satirical pleasantry upon it, if I express a hope, which is earnest and true, that, if there be an immortality for any class of dumb animals, the donkey shall go into it first, and have a better place in it than their parlor dogs or nicely-groomed horses. Evidently they are building up a claim to this illustrious distinction of another existence for these pets on the sole ground of merit, not of works, even, but of mere intelligence, fidelity, and affection. Granted; but the donkey should go in first and take the highest place on that basis. When you come to the standard of moral measurement, it may be claimed as among the highest of human as well as animal virtues, "to learn to suffer and be strong." And this virtue the donkey has learned and practised incomparably beyond any other creature that ever walked on four legs since the Flood. Let these good people remember that their fanciful and romantic favoritisms are not to rule in the destinies awarded to the infinitesimally human spirits of domestic animals in another world, if another be in reserve for them. Let them remember that their softly-cushioned dogs, and horses so delicately clad, and fed, and fondled, have had a pretty good time of it in this life, and that in another, the poor, despised, abused donkey, going about begging, with such a long and melancholy face, for withered cabbage leaves and woody-grained turnips cast out and trodden under feet of happier animals,—that this meek little creature, kicked, cuffed, and club-beaten all the way from hopeless youth to an ignominious grave, will carry into another world merits and mementoes of his earthly lot that will obtain, if not entitle him to, some compensation in the award of a future condition. It is treading on delicate ground even to set one foot within the pale of their unscriptural theory; but as many of them hold the Christian faith in pureness of living and doctrine, let me remind them of that parable which shows so impressively how the dispar-

ities in human condition here are reversed in the destinies of the great hereafter.

But, to return to the earthly lot and position of this poor, libelled animal. Among all the four-footed creatures domesticated to the service of man, this has always been the veriest scapegoat and victim of the cruellest and crabbedest of human dispositions. Truly, it has ever been born unto sorrow, bearing all its life long a weight of abuse and contumely which would break the heart of a less sensitive animal in a single week. From the beginning it has been the poor man's beast of burden; and "pity 'tis 'tis true," poor men, in all the generations of human poverty, have been far too prone to harshness of temper and treatment towards the beasts that serve them and share their lot of humble life. The donkey is made a kind of Ishmaelite in the great family of domestic animals. He is made, not born so. He is beaten about the head unmercifully with a heavy stick, and then jeered at for being stupid and obstinate! just as if any other creature, of four or two legs, would not be stupid after such fierce congestion of the brain. His long ears subject him to a more cruel prejudice than ever color engendered in the circle of humanity but just above him. True, he is rather unsymmetrical in form. His head is disproportionately long and large, quite sufficient in these dimensions to fit a camel. He is generally a hollow-backed, pot-bellied creature, about the size of a yearling calf, with ungainly, sloping haunches, and long, coarse hair. But nearly all these deformities come out of the shameful treatment he gets. You occasionally meet one that might hold up its head in any animal society; with straight back, symmetrical body and limbs, and hair as soft and sleek as the fur of a Maltese cat; with contented face, and hopeful and happy eyes, showing that he has a kind master.

The donkey is really a useful and valuable animal, which might be introduced into America with great advantage to our farmers. I know of no animal of its size so tough and strong. It is astonishing, as well as shocking, to see what loads he is made to draw here. The vehicle to which he is usually harnessed is a heavy, solid affair, frequently as large as our common horse-carts. He is put to all kinds of work, and is almost exclusively the poor man's beast of burden and travel. In cities and large towns, his cart is loaded with the infinitely-varied wares of street trade; with cabbages, fish, fruit, or with some of the thousand-and-one nicknacks that find a market among the masses of the common people. At watering-places, or on the "commons" or suburban playgrounds of large towns, he is brought out in a handsome saddle, or a well got-up little carriage, and let by the hour or by the ride to invalid adults, or to children bubbling over with life. Here, although the everlasting club, to which he is born, is wielded by his driver, he often looks comfortable and sleek, and sometimes wears a red ribbon at each ear. It would not pay to bring on to the ground the scrawny, bony creature that generally tugs in the costermonger's cart. It is in the coal region or trade that you meet with him and his driver in their worst apostacy from all that is seemly in man or beast. To watch the poor creature, begrimed with coal-dust, wriggling up a long, steep hill, with a load four times his own weight, griping with his little sheep-footed hoofs into the black, slimy pavement of the road, while his tall, sooty-faced and harsh-voiced master, perhaps sitting on the top or on a shaft, is punching and beating him; to see this is enough to stir up the old adam in the meekest Christian to emotions of pugilistic indignation. It has often cost me a doubtful and protracted effort to keep it down. Indeed, I have often yielded to it so far as to wish that once more the poor creature might be honored of God with His gift to Balaam's ass, and be able to speak, bolt outright, an indignant remonstrance, in human speech, against such treatment. It would serve them right!—these lineal descendants of Balaam, who have inherited his club and wield it more cruelly.

A word or two more about this animal, and I will pass on to others of more dignity of position. He is the cheapest as well as smallest beast of burden to be found in Christendom. You may buy one here for twenty or thirty English shillings. I am confident that they would be extremely serviceable in America, if once introduced. It costs but very little to keep them, and they will do all kinds of work up to the draught of 600 or 800 lbs. You frequently see here a span of them trotting off in a cart, with brisk and even step. Sometimes they are put on as leaders to a team of horses. I once saw on my walk a heavy Lincolnshire horse in the shafts, a pony next, and a donkey at the head, making a team graduated from 18 hands to 6 in height; and all pulling evenly, and apparently keeping step with each other, notwithstanding the disparity in the length of their legs.

It would be unjust to that goodwill to man and beast which is being organised and stimulated in England through an infinite number of societies, if I should omit to state that, at last, a little rill of this benevolence has reached the donkey. That most valuable and widely-circulated penny magazine, "The British Workman," and its little companion for British workmen's children, "The Band of Hope Review," have advocated the rights and better treatment of this humble domestic for several years. His cause has also been pleaded in a packet of little papers called "Leaflets of the Law of Kindness for the Children." And now, at last, a wealthy and benevolent champion, on whom the mantle of Elizabeth Fry, his aunt, has fallen, has taken the lead in the work of raising the useful creature to the level of the other animals of the pasture, stable, and barn-yard. Up to the present time, every creature that walks on four or two legs, either haired, woolled, or feathered, with the single exception of the donkey has had the door of the Agricultural Exhibition thrown wide open to it, to enter the lists for prizes or "honorable mention," and for general admiration. A pig, whose legs and eyes have all been absorbed out of sight by an immense rotundity of fat, is often decked with a ribbon, of the Order of the Garter genus, as a reward of merit, or of grace of form and propor-

tions! Turkeys, geese, ducks, and hens of different breeds, strut or waddle off with similar distinctions. As for blood-horses, bulls, cows, and sheep, one not versed in such matters might be tempted to think that men, especially the poorer sort, were made for beasts, and not beasts for men. And yet, *mirabile dictu!* at these great social gatherings of man-and-animal kind, there has not been even "a negro-pew" for the donkey. A genuine, raw, Guinea negro might have as well entered the Prince of Wales' Ball in New York bare-footed, and offered to play a voluntary on his banjo for the dancers, as this despised quadruped have hoped to obtain the *entree* to these grand and fashionable assemblies of the shorter-eared *elite* of society.

But this prejudice against color and long ears is now going the way of other barbarisms. The gentleman to whom I have referred, a Member of Parliament, whose means are as large as his benevolence, has taken the first and decisive step towards raising the donkey to his true place in society. He has offered a liberal prize for the best conditioned one exhibited at the next Agricultural Fair. Since this offer was made, a very decided improvement has been noticed among the donkeys of the London costermongers, as if the competition for the first prize was to be a very large one.

It will be a kind of St. Crispin's Day to the whole of the long-eared race—a day of emancipation from forty centuries of obloquy and oppression. Doubtless they will be admitted hereafter to the Royal Agricultural Society's exhibitions, to compete for honors with animals that have hitherto spurned such association with contempt.

CHAPTER VI.

HOSPITALITIES OF "FRIENDS"—HARVEST ASPECTS—ENGLISH COUNTRY INNS; THEIR APPEARANCE, NAMES, AND DISTINCTIVE CHARACTERISTICS—THE LANDLA-

DY, WAITER, CHAMBERMAID, AND BOOTS—EXTRA FEES AND EXTRA COMFORTS.

I reached Saffron Walden at 4 p.m., notwithstanding my involuntary walk of six extra miles in the morning. Here I remained over the Sabbath, again enjoying the hospitality of a Friend. And perhaps I may say it here and now with as much propriety as at any other time and place, that few persons, outside the pale of that society, have more frequently or fully enjoyed that hospitality than myself. This pleasant experience has covered the space of more than sixteen years. During this period, with the exception of short intervals, I have been occupied with movements which the Friends in England have always regarded with especial sympathy. This connection has brought me into acquaintance with members of the society in almost every town in Great Britain in which they reside; and in more than a hundred of their homes I have been received as a guest with a kindness which will make to my life's end one of its sunniest memories.

On the following Monday, I resumed my walk northward, after a carriage ride which a Friend kindly gave me for a few miles on the way. Passed through a pre-eminently grain-producing district. Apparently full three-fourths of the land were covered with wheat, barley, oats, and beans. The fields of each were larger than I had noticed before; some containing 100 acres. The coming harvest is putting forth the full glory of its golden promise. The weather is all a farmer could wish, beautiful, warm, and bright. Nature, in every feature of its various scapes, seems to smile with the joy of that human happiness which her ministries inspire. Here, in these still expanses, waving with luxuriant crops, apparently so thinly peopled, one, forgetting the immense populations crowded into city spaces, is almost tempted to ask, where are all the mouths to eat this wide sea of food for man and beast, softening so gently into a yellow sheen under the very rim of the distant horizon? But, in the great heart of London, beating with the wants of millions, he will be likely to reverse the question, and ask, where can one buy bread wherewith to feed this great multitude?

At Sawston, a rustic little village on the southern border of Cambridgeshire, I entered upon the enjoyment of English country-inn life with that relish which no one born in a foreign land can so fully feel as an American. As one looks upon the living face of some distinguished celebrity for the first time, after having had his portrait hung up in the parlor for twenty years, so an American looks, for the first time, at that great and picturesque speciality among human institutions, the village inn of old England. The like of it he never saw in his own country and never will. In fact, he would not like to see it there, plucked up out of its ancient histories and associations. In the ever-green foliage of these it stands inwoven, as with its own network of ivy. Other countries, even older than England, have had their taverns from time immemorial; but they are all kept in the background of human life. They do not come out in contemporaneous history with any definiteness; not even accidentally. If a king is murdered in one of them, or if it is the theatre of the most thrilling romance of love, you do not know whether it is a building of stone, brick, or wood; whether it is one, two, or three stories in height. No outlines nor aspects are given you to help to fill up a rational picture of it. Neither the landlord nor the landlady is drawn as a representative man or woman. Either might be mistaken for a guest in their own house, if seen in hat or bonnet by a stranger.

But not so of the English country inn. It comes out into the foreground of a thousand interesting histories and pictures of common life. In them it has an individuality as marked as the parish church, *couchante* in its wide-rimmed nest of grave stones; as marked in unique architecture, location, and surroundings. In none of these features will you find two alike, if you travel from one end of the country to the other; especially among those a century old. You might as well mistake one of the living animals for the other, as to mistake "The Blue Boar" for "The Red Lion." They differ as much from each other in general make and aspect as do their nominal prototypes. To give every one of their thousands "a local habitation and a name" of striking distinctness, has required an ingenuity which has produced many interesting feats of house-building and nomenclature. Both these departments of genius figure largely in the poetry and classics of the institution, with which the reading million of America have been familiar from youth up. And when any of them come to travel in England, it will greatly enhance their enjoyment to find that the pictures they have admired and the descriptions they have read of the famous country inn have been true to the very life and letter. All its salient features they recognise at once, and are ready to exclaim, "How natural!" meaning by that, how true is the original to the picture which they have seen so frequently. If they go far enough, they will find the very original of every one of the hundred pictures they have seen, painted by pen or pencil. They will find that all of them have been true copies from nature. Here is the portly-looking, well-to-do, two-story tavern, standing out with its comfortable, cream-colored face broadside to the street. It is represented in the old engraving with a coach-and-four drawn up before the door, surrounded by a crowd of spectators and passengers, some descending and ascending on ladders over the forward wheels; some looking with admiration at the scarlet coats of the pursy and consequential driver and guard; some exchanging greetings, others farewell salutations; ostlers in long waistcoats, plush or fustian shorts, and yellow leggings, standing bareheaded with watering-pails at the "'osses' 'eads;" trunks great and small going up and down; village boys in high excitement; village grandfathers looking very animated; the landlord, burly, bland, and happy, with a face as rotund and genial as the full moon shining upon the scene; and those round, rosy, sunny, laughing faces peering out of the windows with delightful wonderment and exhilaration, winked at by the driver, and saluted with a graceful motion of his whip-handle in recognition of the barmaid, chambermaid, and all the other maids of the house. The coach, with all its picturesque appointments, its four-in-hand, the stirring heraldry of its horn coming down the road, its rattling wheels, the life and stir aroused and moved in its wake,—all this has gone from the presence of a higher civilisation. It will never re-appear in future pictures of actual life in England. It is all gone where the hedges and hedge-row trees will probably go in their turn. But the same village inn remains, and can be as easily recognised as a widow in weeds, who still wears a hopeful face, and makes the best of her bereavement.

But that humbler type of hostelry so often represented in sketches of English rural life and scenery—the little, cozy, one-story, wayside, or hamlet inn, with its thatched roof, checker-work window, low door, and with a loaded hay-cart standing in front of it, while the driver, in his round, wool hat, and in his smock-frock, is drinking at a pewter mug of beer, with one hand on his horse's neck—this the hand of modern improvements has not yet reached. This may be found still in a thousand villages and hamlets, surrounded with all its rural associations; the green, the geese, and gray donkeys feeding side by side; low-jointed cottages, with long, sloping roofs greened over with moss or grass, and other objects usually shadowed dimly in the background of the picture. It is these quiet hamlets and houses in the still depths of the country, away from the noise and bluster of railway life and motion, that best represent and perpetuate the primeval characteristics of a nation. These the American traveller will find invested with all the old charm with which his fancy clothed them. It will well repay him for a month's walk to see and enjoy them thoroughly.

In these days of sun-literature, whose letters are human faces, and whose new volumes are numbered by the million yearly, without a duplicate to one of them, I am confident that a volume of these English village inns of the olden school, in photographs, would command a large sale and admiration in America, merely as specimens of unique and interesting architecture. A thousand might be taken, every one as unlike the other in distinctive form and feature, as every one of the same number of men would be to the other.

The diversification of names, being more difficult, is still more remarkable. Although the spread eagle figures largely as the patron genius of American hotels, still nine-tenths of them bear the names of states, counties, towns, or national or local celebrities. But here natural history comes out strong and wide. The heraldry of sovereigns, aristocracy, gentry, commercial and industrial interests, puts up its various *arms* upon hundreds of inns in town and country. All occupations and recreations are well represented. Thus no country in the world approaches England in the wide scope and play of hotel nomenclature. Some of the combinations are exceedingly unique and most interesting in their incongruity. Dickens has not exaggerated this characteristic; not even done it justice in his hotel scenes. Things are put together on a hundred tavern signs that were never joined before in the natural or moral world, and put together frequently in most grotesque association. For instance, there is a large, first-class inn right in the very heart of London, which has for a sign, not painted on a board, but let into the wall of the upper story, in solid statuary, a huge human mouth opened to its utmost capacity, and a bull, round and plump, standing stoutly on its four legs between the two distended jaws. Now, the leading idea of this device is involved in a tempting obscurity, which leads one, at first sight, into different lines of conjecture. What did the designer of this group of statuary really intend to represent? Was it to let the outside world know that, in that inn, the "Roast Beef of Old England" was always to be found *par excellence*? If so, would a man's mouth swallowing a bull whole, and apparently alive, with hide and horns, tend to stimulate the appetite of a passing traveller, and to draw

him into the establishment? But leaving these ambiguous symbols to be interpreted by the passing public according to different perceptions of their meaning, how many in a thousand would guess aright the name given to the tavern by these tokens? Would not ninety-nine in a hundred say, "The Mouth and Bull," to be sure, not only on the principle that the major includes the minor, but also because the human element is entitled to precedence in the picture? But the ninety-nine would be completely mistaken, if they adopted this natural conclusion. They would find they had counted without their host, who knows better than they the relative position and value of things. What has the law of logic to do with fat beef? The name of his famous hotel is "THE BULL AND MOUTH;" and few in London have attained to its celebrity as a historical building. One is apt to wonder if this precedence given to the beast is really incidental, or adopted to give euphony to the name of an inn, or whether there is a latent and spontaneous leaning to such a method of association, from some cause or other connected with perceptions of personal comfort afforded at such establishments. Accidental or intentional, this form of association is very common. There is no tavern in London better known than *The Elephant and Castle*, a designation that would sound equally well if the two substantives were transposed. Even the loftiest symbols of sovereignty often occupy the secondary place in these compound titles. There are, doubtless, a hundred inns in Great Britain bearing the name of *The Rose and Crown*, but not one, to my knowledge, called "The Crown and Rose." The same order obtains in sporting sections and terminology. It is always "The *Hare* and Hounds;" never "*Hounds* and Hare."

This characteristic in itself is very interesting, and no American, with an eye to the unique, would like to see it changed. But if the more syntax of hotel names in England is so pleasant for him to study, how much more admirable is their variety! He has read at home of many of them in lively romance and grave history but he finds here that not half has been told him. He is familiar with the Lions, Red, White, and Black; the Bulls and Boars of the same colors; the Black and White Swans and Harts; the Crown and Anchor, the Royal George, Queen's Head, and a few others of similar designation. These names have figured in volumes of English literature which he has perused. But let him travel on the turnpike road through country towns and villages, and he will meet with names he never thought of before, mounted over the doors of some of the most comfortable and delightful houses of entertainment for man and beast that can be found in the world. Here are a few that I have noticed: "The Three Jolly Butchers," "The Old Mash Tub," "The Old Mermaid," "The Old Malt Shovel," "The Chequers," "The Dog-in-Doublet," "Bishop Boniface," "The Spotted Cow," "The Green Dragon," "The Three Horseshoes," "The Bird-in-Hand," "The Spare Rib," "The Old Cock," "Pop goes the Weasel." There are wide spaces between these names which may be filled up from actual life with numbers of equal uniqueness. But it is not in architecture nor in name that the country inn presents its most attractive characteristic. These features merely specialise its outward corporeity. The living, brightening, all-pervading soul of the establishment is the LANDLADY. Let her name be written in capitals evermore. There is nothing so naturally, speakingly, and gloriously English in the wide world as she. It is doubtful if the nation is aware of this, but it is the fact. Her English individuality stands out *embonpoint*, rosy, genial, self-complacent, calm, serene, happyfying, and happy. She is the man and master of the house. She permeates it with her rayful presence, and fills it with a pleasant morning in foggy and blue-spirited days. She it is who greets the coming and speeds the parting guest with a grace which suns, with equal light and warmth, both remembrance and anticipation. It is not put on like a Sunday dress; it is not a thin gloss of French politeness that a feather, blown the wrong way, will brush off. It is not a

color; it is a quality. You see it breathe and move in her like a nature, not as an art. Let no American traveller fancy he has seen England if he has not seen the Landlady of the village inn. If he has to miss one, he had better give up his visit to the Crystal Palace, Stratford-upon-Avon, Abbottsford, or even the House of Lords, or Windsor itself. Neither is so perfectly and exclusively English as the mistress of "The Brindled Cow," in one of the rural counties of the kingdom.

It would be necessary to coin a new word if one were sought to contain and convey the distinctive characteristic of inn-life in England. Perhaps *homefulness* would do this best, as it would more fully than any other term describe the coziness, quiet, and comfort to be enjoyed at these places of entertainment. Not one in a hundred of them ever heard the sound of the hotel-going bell, as we hear it in America. You are not thundered up or down by a vociferous gong. Then there is no marching nor counter-marching of a long line of waiters in white jackets around the dinner table, laying down plate, knife, fork, and spoon with uniform step and motion, as if going through a dress-parade or a military drill. There is no bustle, no noise, no eager nor anxious look of served or servants. Every one is calm, collected, and comfortable. "The cares that infest the day" do not ride into the presence of that roast beef and plum pudding on the wrinkles of any man's forehead, however business affairs may go with him outside. No one is in a hurry to sit down or to arise from the table. The whole economy of the establishment is to make you as much at home as possible; to individualise you, as far as it can be done, in every department of personal comfort. You follow your own time and inclination, and eat and drink when and how you please, with others or alone. The congregate system is the exception, not the rule. It seldom ever obtains at breakfast or tea. In many cases you have a little round table all to yourself at these meals. But if there is a common table for half a dozen persons, the tea and toast and other eatables are never aggregated into a common stock.

Each person if he is a single guest, has his own allotment, even to a separate tea-pot. The *table d'hote*, if there be one at all, is made up like a select dinner party, rather early in the morning. If the guests of the house are not directly invited, they are asked, in a tone of hospitality, if they will join in the social meal, the only one got up by the establishment at which the table is not mapped out in separate holdings, or little independencies of dishes, each bounded by the wants and capacities of the individual occupant.

The presiding and working faculty of a common English inn distinguishes it by another salient characteristic from the hotels of other countries. The landlady is, of course, the president of the establishment, whether or not she calls any man lord in the retired and family department of the house. But the actual *gerantes*, or working corps, with which you have to do immediately, are three independent and distinct personages, called the waiter, chambermaid, and *boots*. If it were respectful to gender, these might be called the great triumvirate of the English inn. No traveller after a night's lodging and breakfast, will mistake or confound the prerogatives or perquisites of these officials. If he is an American, and it be his first experience of the *regime*, he will be surprised and puzzled at the *imperium in imperio* which his bill, presented to him on a tea-tray, seems to represent. In no other business transaction of his life did he ever see the like. It goes far beyond anything in the line of limited partnership he ever saw. There is only one partial parallel that approaches it; and this comes to his mind as he reads the several items on his bill. When made out and interpreted, it comes to this: the proprietor, the waiter, chambermaid, and boots are independent parties, who get up a night's lodging and two or three meals for you on the same footing as four independent underwriters would take proportionate risks at Lloyd's in some ship at sea. Or, what would put it in simpler form to an uninitiated guest, he is apparently first charged for the raw provisions he consumes, and for the rent of his bed-room. This is the proprietor's share. Then, there is a separate charge for each of the remaining items of the entertainment,—for cooking and serving up each meal, for making up your bed, and for blacking your boots; just as distinctly as if you had gone out into the town the previous evening and hired three separate individuals to perform these services for you; and as if you had no right nor reason to expect from the landlord a dinner all cooked and served, but that you only bought it in the larder.

Now, this is a peculiarity of the English hotel system that is apt to embarrass travellers from other countries, especially from America, where no such custom could be introduced. I do not know how old the custom is in Great Britain. Doubtless it originated in the almost universal disposition and habit of Englishmen of dropping gratuities or charity-gifts here and there with liberal hand, either to obtain or reward extra service in matters of personal comfort, or to alleviate some case of actual or stimulated suffering that meets them. It was natural and inevitable that gratuities thus given to hotel servants frequently to stimulate and reward special attention should soon become a rule, acting upon guests like a law of honor. When so many gave, and when the servants of every hotel expected a gift, a man must feel shabby to go away without dropping a few pennies into the hands of eager expectants who almost claimed the gratuity as a right. The worst stage of the system was when the expected gift was measured by your supposed position and ability, or when the waiter or the chambermaid, flattering you with what Falstaff would call an instinctive perception of your dignity, would say with an asking and hopeful smile, "What you please, sir." Now, that was not the question with you at all. You wanted to know how much each expected, or how much you must give to acquit yourself of the charge of being "a screw," when they put their heads and gains together in conference and comparison after you were gone. So, on the whole, it was a great relief when all

these awkward uncertainties of expectation were cleared up and rectified in the system now usually adopted.

Whether you be rich or poor, or whatever position or pretension be attributed to you, the fees of the universal triumvirate are put down specifically in black and white among the other charges on your bill. As I hope these notes may convey some useful information to Americans who may be about to visit England for the first time, it may be of some use to them to state what is the usual rule in this matter at the middle-class hotels in this country; for with those of the first rank I never have made nor ever expect to make any personal acquaintance. A moderate bill for a day's entertainment will read thus:—

	s.d.
Tea (bread and butter or toast)	1 0
Bed	1 6
Breakfast (rasher of bacon, eggs, or cold meats)	1 6
Dinner	2 6
Waiter	0 9
Chambermaid	0 6
Boots	0 3

Total	8 0

These are about the average charges at the middle-class hotels in Great Britain. Generally the servants' fees amount to 25 per cent. of the whole bill. These, too, are graduated to parts of days. The waiter expects 3d. for every meal he serves; the chambermaid 6d. for every bed she makes, and the boots 3d. for doing every pair of boots, brogans, or shoes. You will pay these charges with all the better grace and good-will to these servants when you come to learn that these fees frequently, if not always, constitute all the salary they receive for hotel service. Even in a great number of eating-shops the same rule obtains. The penny you give the waiter, male or female, is all he or she gets for serving you. Besides this consideration, you get back much additional personal comfort from these extras. The waiter serves you with extra satisfaction and assiduity un-

der their stimulus. He acts the host very blandly. He answers a hundred questions, extraneous to the meal, with good-natured readiness. He is a good judge of the weather and its signs. He is well "posted-up" in the local histories and sceneries of the place. He can give political information on both sides, incidents and anecdotes to match, whether you are Tory, Whig, or Radical. If you have a bias in that direction, he has or has heard some thoughts on Bishop Colenso and the Tractarians. In short, he caters to the humour and disposition of every guest with a happy facility of adaptation; and the shilling you give him at the end of a day's entertainment has been pretty well earned, if you have availed yourself of all these extra attentions which he is prepared and expecting to give for it.

The same may be said of the chambermaid. She is not the taciturn invisible that steals in and out of your bed-room, and does it up when you are at breakfast or at your out-door business—whom you never see, except by sheer accident, as in the American hotel. She is an important and prominent personage in the English inn. She is a kind of mistress of the robes, and exercises her prerogative with much conscious dignity and self-satisfaction; and, what is better, with great satisfaction to yourself. No other subordinate official or servant trenches or poaches upon her preserves. She it is who precedes you up stairs with a candle, on a broad-bottomed brass candlestick, polished to its highest lustre. She conducts you to your room as if you were her personal guest, invited and expected a month ago. She opens the door with amiable complacency, as if welcoming you to a hospitality which she had prepared for you with especial care, before she knew you had arrived in town. She invites you, by a movement of her eyes, to glance at the room and see how comfortable it is; how round and soft is the bed, how white and well-aired are the sheets and pillows, how nice the curtains, how clean and tidy the carpet, in short, how everything is fitted to incline you to "rest and be thankful." And then the cheery "*good night!*" she bids you is said with a tone that is worth the sixpence she expects in the morning; and you pay it, too, with a much better grace than could be expected from an American recently arrived in the country.

And the "boots" is a character, too, unmixedly and interestingly English, in name, person, appearance, and position. In the first of these qualities he is unique, being called after the subject of his occupation. He is an important personage, and generally has his own bell in the dining-room, surmounted by his name, to be called for any service coming within his department. And this is quite a wide one, including a great variety of errandry and porterage, as well as polishing boots and shoes. He is very helpful in a great many different ways, and often very intelligent, and knows all about the streets, the railway trains, the omnibuses, cabs, etc., and will assist you in such matters with good grace and activity. He may have got in the way of putting the H before the eggs instead of the ham; but he is just as good for all that, and more interesting besides. So you do not grudge the 3d. you give him daily for his strictly professional services, or the extra 6d. he expects for carrying your carpet-bag or portmanteau to the railway-station.

Thus, although this feeing of servants may seem at first strange to an American traveller in England, and may occasion him some perplexity and even annoyance, he will soon become accustomed to it; and in making up the balance-sheet of the additional cost on one side and the additional comfort on the other which the system produces, he will come even to the mathematical conclusion, "if to equals you add equals, the sums will be equals."

CHAPTER VII.

LIGHT OF HUMAN LIVES—PHO-TOGRAPHS AND BIOGRAPHS—THE LATE JONAS WEBB, HIS LIFE, LABORS, AND MEMORY.

The next morning I resumed my walk and visited a locality bearing a name and association of world-wide celebrity and interest. It is the name of a small rural hamlet, hardly large enough to be called a village, and marked by no trait of nature or art to give it distinction.

There are conditions and characteristics both in the natural and moral world which can hardly be described fully in Saxon, Latin, or Greek terminology, even with the largest license of construction. There are attributes or qualities attaching to certain locations, of the simplest natural features, which cannot even be hinted at or suggested by the terms, *geography, topography, or biography*. Put the three together and condense or collocate their several meanings in one compound qualification which you can write and another spell, and you do not compass the signification you want to convey. The soul of man has its immortality, and the feeblest-minded peasant believes he shall wear it through the ages of the great hereafter. The literature of human thoughts claims a life that shall endure as long as the future existence of humanity. The memory of many human actions and lives puts in a plea and promise of a duration that shall distance the sun's, and overlap upon the bright centuries of eternity. The human body, even, is promised its resurrection by the divinest authority and illustration, and waits hopefully, under all its pains and weaknesses, for the glory to be revealed in it when the earth on which it dwells shall have become "a forgotten circumstance." Human loves, remembrances, faiths, and fellowships lift up all their meek hands to the Father of Spirits, praying to be lifted up into His great immortality, and to be permitted to take with them unbroken the associations that sweetened this earthly life. Many humble souls that have passed through the furnace of affliction, poverty, and trial seven times heated, and heated daily here, have believed that He who went up through the same suffering to His great White Throne, would let them sing beside the crystal waters the same good old psalm tunes and songs of Sion which they sang under the willows

of this lower world of tears and tribulation. How all the sparks of the undying life in man fly upward to the zenith of this immortality! You may call the steep flights of this faith pleasant and poetical diversions of a fervid imagination, but they are winged with the pinions that angels lift when they soar; pinions less ethereal than theirs, but formed and plumed to beat upward on the Milky Way to their Source, instead of swimming in the thinly-starred cerulean, in which spirits, never touched with the down or dust of human attributes, descend and ascend on their missions to the earth. Who can have the heart to handle harshly these beautiful faiths? To say, this hope may go up, but this must go down to the darkness of annihilation! Was it irreverent in the pious singing-master of a New England village, when he said, that often, while returning home late on bright winter nights, he had dropped the reins upon his horse's neck, and sung Old Hundred from the stars, set as notes to that holy tune, when they first sang together in the morning of the creation? What spiritual good or Christian end would be gained, to break up the charm and cheer of this his belief? Or to dispel that other confidence, which so helped him to bear earth's trials, that one day he should join all the spirits of the just made perfect, and all the high angels in heaven, and, on the plane of that golden gamut, they should sing together their hymns of joy and praise, in that same, good, old tune, from those same star-notes, which a thousand centuries should not deflect nor transpose from their first order within those everlasting staves and bars!

If the spirit's faith be allowed such wide confidences as these; if it may carry up into the invisible and infinite so many precious relics from the wreck of time, so many human circumstances and associations, why may it not take with it, to hang up in its heaven, photographs of those earthly localities rendered immortal here by the lives of good and great men? Such a life is a sun, and it casts a disk of light upon the very earth on which it shines; not that flashy circle which the lens of the microscope casts upon the opposite wall, to show how scarcely visible mites may be magnified; but a soft and steady illumination that does not dim under the beating storms and bleaching dews of centuries, but grows brighter and brighter, as if the seed-rays that made it first multiplied themselves from year to year. The earth becomes more and more thickly dotted with these permanent disks of light, and each is visited by pilgrims, who go and stand with reverence and admiration within the cheering circle. Shakespeare's thought-life threw out a brilliant illumination, of wide circumference, at Stratford-upon-Avon, and no locality in England bears a biograph more venerated than the birthplace of the great poet. His thought-life was a sun that will never set as long as this above us shines. It is rising every year to new generations that never saw its rays before. When he laid down his pen, at the end of his last drama, the whole English-speaking race in both hemispheres did not number twice the present population of London. Now, seventy-five millions, peopling mighty continents, speak the tongue he raised to the grandest of all earth's speeches; and those who people the antipodes claim to offer the best homage to his genius. Thus it will go on to the end of time. As the language he clothed with such power and might shall spread itself over the earth, and be spoken, too, by races born to another tongue, his life-rays will permeate the minds of countless myriads, and the more widely they diverge and the farther they reach, the brighter and warmer will be the glow and the flow of that disk of light that embosoms and illumines his birth-place in England.

What is true of Stratford-upon-Avon is equally true of Abbotsford, of the birth-place of Milton, Burns, Bunyan, Baxter, and other great minds, which have shone each like a sun or star in its sphere. Now, what one word, recognised as legitimate in scientific terminology, would describe fully one of these disks of light cast by a human life upon a certain space of earth, not as a fugitive flash, but as a permanent illumination? *Photograph* would not do it,

because its meaning is fixed and rigidly technical, as simple light-writing, or sun-writing. The term is completely pre-occupied by this signification, and you cannot inject the human life element into it. *Biography* is universally limited to an operation in which the life is the subject, not the agent. It is simply the writing out of a life's history by some one with a common goose-quill or steel pen. Still, the word *biograph* would be the best, of the same length, that we could form to describe one of these disks of light, if it were made the same verb active as *photograph*; or to mean that the life is the agent, as well as the subject,—that it writes itself in light upon a certain locality, just as the sun graves a human face upon glass. Let us then call the bright and quenchless planispheres, which such lives describe and fill around them, *biographs*, assuming that the script is in rays of light. As differ the stars above in glory, so these differ in the qualities of their illumination. The brightest of them, to mere human seeming, are those which shine with the sheer brilliancy of intellect and genius. These chiefly halo the homes of "the grand old masters" of poetry, painting, eloquence, and martial glory. These attract to their disks pilgrims the most numerous and enthusiastic. But, as the nearest stars are brightest, not largest, so these biographs are brightest on their earth-side. There are thousands of less sharp and spangling lustre to the eyes of the multitude, which shine with tenfold more brilliancy from their eternity-face. These are they that halo the homes of good men, whose great hearts drank in the life of God's love in perpetual streams, and distilled it like a luminous dew around them; men whose thoughts were not mere scintillations of genius, but living labors of beneficence, bearing the proof as well as promise of that immortality guaranteed to the deeds of earth's saints. If the soul, after such long isolation, is to take again to its embrace so much of the old human corporeity it wore here below, does it transcend the prerogative of hope in the great resurrection to believe that these biographs of God's loving children on earth shall

be taken up whole into the same immortality as the bodies in which they worked His will among men? Is the faith too fanciful or irreverent that believes, that the corridors and inner temples of Heaven's Glory will be hung with these biographs of His servants surrounding, like stars, the light-flood of His love that radiated from His cross on earth? Is it too presumptuous to think and say, that such pictures will be as precious in His sight as any graven by the lives of angels on their outward or homeward flights of duty and delight? These are they, therefore, that shall give to the earth all the immortality to which it shall attain. These are they that shall take up into the brilliant existence of the hereafter, ten thousand sections of its corporeity; portions of its surface, perhaps, as substantial as the human form that the souls of men shall wear in another world. These are they that shall shine as the stars, when those beaming so brilliantly in our eyes around the shrines of mere intellect and genius, shall have "paled their ineffectual fires" before the efflux of diviner light. Let him, then, of thoughtful and attentive faculties think on these great and holy possibilities, when he treads within the pale of a good man's life, whose labors for human happiness "follow him" according to divine promise; not out of the world, not down into the grave with his resting body, but out among living generations, breathing upon them and through them a blessed and everlasting influence. Let him tread that disk of light reverentially, for it is the holiest place on the earth's surface outside the immediate circumference of Cavalry.

This is Babraham; and here lived Jonas Webb; a good man and true, whose influence and usefulness had a broader circumference than the widest empire in the world. A Frenchman has written the fullest history of both, and an American here offers reverentially a tribute to his worth. The light of his life was a soft and gentle illumination on its earth-side; the lustre of the other was revealed only by partial glimpses to those who leaned closest to him in the testing-moments of his higher nature. He was one of the great benefactors, whose lives and labors become the common inheritance of mankind, and whose names go down through long generations with a pleasant memory. To a certain extent, he was to the great primeval industry of the world, what Arkwright, Watts, Stephenson, Fulton and Morse were each to the mechanical and scientific activities of the age. He did as much, perhaps, as any man that ever preceded him, to honor that industry, and lift it up to the level of the first occupations of modern times, which had claimed higher qualities of intelligence, genius and enterprize. He was a farmer, and his ancestors had been farmers from time immemorial. He did not bound into the occupation as an enthusiastic amateur, who had acquired a large fortune by manufacturing or commercial enterprize, which he was eager to lavish upon bold and uncertain experiments. He attained his highest eminence by the careful gradations of a continuous experience, reaching back far into the labors of his ancestors. The science, skill and judgment he brought to bear upon his operations, came from his reading, thinking, observations and experiments as a practical and hereditary farmer. The capital he employed in expanding these operations to their culminating magnitude, he acquired by farming. The mental culture, the generous dispositions, the refined manners, the graceful and manly bearing which made him one of the first gentlemen of the age, he acquired as a farmer. The mansion which welcomed to its easy and large-hearted hospitalities guests of such distinction from his own and other countries, was a farmer's home, and few ever opened their doors to more urbanity and cordial cheer. This is an aspect of his character which all those who follow the profession he honored should admire with a laudable *esprit de corps*.

As a back-ground is an important element in the portraiture of human forms or natural scenery, so the ground on which the life and labors of Jonas Webb should be sketched, merits a few preliminary lines. Of all the occupations that employ and sustain the toiling myriads of our race, agriculture leans closest to the bosom of Divine Providence. It is an industry bound to the great and beautiful economics of the creation by more visible and sensible ties than any other worked by human hands. We will not here diverge to dwell upon these high and interesting affiliations. In their place we will give them a little extended thought. There is one feature of agricultural enterprize, however, that should not be overlooked in this connection. All its operations are above-board and open to the wide world, just like the fields to which they are applied. Nothing here is under lock and key. Nothing bears the grim warning over the bolted door, "No admittance here except on business!"—meaning by business, exclusively and sharply, the buying of certain wares of the establishment at a good round profit to the manufacturer, without carrying away a single scintillation or suggestion of his skill. If he has invented or adopted machinery or a process of labor which enables him to turn out cheap muslin at three farthings' less cost per yard than his neighbors can make it, he seals up the secret from them with the keenest vigilance. Not so in the great and heaven-honored industry of agriculture. Its experiments and improvements upon the earth's face are all put into the common stock of human knowledge and happiness. They can no more be placed under lock and key, as selfish secrets, than the stars themselves that look down upon them with all their golden eyes. No new implement of husbandry, no new mechanical force or chemical principle, no new process of labor or line of economy is withheld from the great commonwealth of mankind. As the broad skies above, as the sun and moon, and stars, as the winds, the rains, the dews, the birds and bees of heaven over-ride and ignore, in their missions, the boundaries of jealous nations, so all the great activities of agriculture prove their lineage by following the same generous rule. They are bounded by no nationalities. They are shut up in no narrow enclosure of self, but are put out as new vesicles of light to brighten the general illumination of the world.

The department in which Jonas Webb attained to his position and capacity of usefulness was peculiarly marked by this characteristic. In a certain sense, it occupied a higher range of interest than that section of agriculture which is connected solely with the growing of grain, grass, and other crops. His great and distinguishing husbandry was the cultivation of animal life. To make two spires of grass grow where only one grew before has been pronounced as a great benefaction; and greater still are the merit and the gain of making one grow where nothing grew before. To go into the midst of Dartmoor, and turn an acre of its cold, stony, water-soaked waste into a fruitful field of golden grain, is going into co-partnership with Providence in the work of creation to a very large and honored degree. But to put the skilful hand of science upon creatures of flesh and blood, to re-form their physical structures and shapes, to add new inches to their stature, straighten their backs, expand their reins, amplify their chests, reduce all the lines and curves of their forms to an unborn symmetry, and then to give silky softness and texture to their aboriginal clothing—this seems to be mounting one step higher in the attainment and dignity of creative faculties. And this pre-eminently was the department in which Jonas Webb acquired a distinction perhaps unparalleled to the present time. This has made his name familiar all over Christendom, and honored among the world's benefactors. Never, before him, did a farm-stead become such a centre and have such a wide-sweeping radius as his. None ever possessed such centripetal attractions, or exerted such centrifugal influences for the material well-being of different and distant countries. Indeed, those most remote are most specially indebted to his large and generous operations. America and Australia will ever owe his memory an everlasting homage.

His operations filled and crowned two great departments of improvement seldom, if ever, carried on simultaneously and evenly to a great success by one man. His first distinguishing speciality was sheep-culture. When he had brought this to the highest standard of perfection ever attained, he devoted the surplus capital of skill, experience and pecuniary means he had acquired from the process to the breeding of cattle; and he became nearly as eminent in this field of improvement as in the other. A few facts may serve as an outline of his progress in both to the American reader who is familiar with the general result of his efforts.

Jonas Webb was born at Great Thurlow, Suffolk, on the 10th of November, 1796. His father, who died at the age of ninety-three, was a veteran in agriculture, and had attained to honorable distinction by his efforts to improve the old Norfolk breed of sheep, and by his experiments with other races. The results obtained from these operations convinced his son that more mutton and better wool could be made per acre from the Southdown than from any other breed, upon nine-tenths of the arable land of England, where the sheep are regularly folded, especially where the land is poor. In 1822, he commenced that agricultural career which won for him such a world-wide celebrity, by taking the Babraham Farm, occupying about 1,000 acres, some twelve miles south of Cambridge. In a very interesting letter, addressed to the *Farmers' Magazine*, about twenty years since, he gives a valuable *resume* of his experience up to that time. In this he states several facts that may be especially useful to American agriculturists. Having decided in his own mind that the Southdowns were preferable to every other breed, for the two properties mentioned, he went into Sussex, their native county, and purchased the best rams and ewes that could be obtained of the principal breeders, regardless of expense, and never made a cross from any other breed afterwards. Nor was this all; he never introduced new blood into his stock from flocks of the same breed, but, by a virtually in-and-in process, he was able to produce qualities till then unknown to the race, and to make them permanent and distinctive properties.

Now this achievement in itself has an interest beyond its utilitarian value to the agricultural world. To

"Rejoice in the joy of well-created things"

is one of the best privileges and pleasures of a well-constituted mind. But what higher honor can attach to human science or industry than that of taking such a visible and effective part in that creation?—in sending out into the world successive generations of animal life, bearing each, through future ages and distant countries, the shaping impress of human fingers, long since gone back to their dust; features, forms, lines, curves, qualities and characteristics which those fingers, working, as it were, on the right wrist of Divine Providence, gave to the sheep and cattle upon a thousand hills in both hemispheres? There are flocks and herds now grazing upon the boundless prairies of America, the vast plains of Australia, the steppes of Russia, as well as on the smaller and greener pastures of England, France, and Germany, that bear these fingermarks of Jonas Webb, as mindless but everlasting memories to his worth. If the owners of these "well-created things" value the joy and profit which they thus derive from his long and laborious years of devotion to their interests, let them see that these finger-prints of his be not obliterated by their neglect, but be perpetuated for ever, both for their own good and for an ever-living memorial to his name.

It is a fact of instructive suggestion, that although Mr. Webb commenced his operations in 1822, he won his first prize for stock ewes at the meeting of the Royal Agricultural Society at Cambridge in 1840. Here he realised one of the serious disadvantages to which stock-breeders in England are exposed, in "showing" sheep, cattle or swine at these annual exhibitions. The great outside world, with tastes that lean more to fat sirloins or shoulders than to the better symmetries of animated nature, almost demands that every one of these unfortunate beasts should be offered up as a bloated, blowing sacrifice to those great twin idols of fleshy lust, Tallow

and Lard. If, therefore, a stock-raiser has not decided to drive his Shorthorn cow or Southdown ewe immediately from the Fair-grounds to the butcher's shambles, he runs an imminent risk of losing entirely the use and value of the animal. So great is this risk, that much of the stock that would be most useful for exhibition is withheld, and can only be seen by visiting private establishments scattered over the kingdom. They are too valuable to run the terrible gauntlet of oil-cake, bean and barley-meal, through which they must flounder on in cruel obesity to the prize. Especially is this the case with breeding animals. Mr. Webb's experience at his first trial of the process, will illustrate its tendencies and results. Of the nine shearling ewes he "fed" for the Cambridge Show, he lost *four*, and only raised two or three lambs from the rest. At the Exhibition of 1841, at Liverpool, he won three out of four of the prizes offered by the Royal Agricultural Society for Southdowns, or any other short-woolled sheep; two out of four offered at Bristol, in 1842, and three out of four offered at Derby, in 1843. But here again he over-fed two of his best sheep, under the inexorable rule of fat, which exercises such despotic sway over these annual competitions, and was obliged to kill them before the show. It will suffice to show the loss he incurred by this costly homage to Tallow, to give his own words on the subject:—"I had refused 180 guineas for the hire of the two sheep for the season. I also quite destroyed the usefulness of two other aged sheep by over-feeding them last year. Neither of them propogated [sic] through the season, and I have had each of them killed in consequence, which has so completely tired me of overfeeding that I never intend exhibiting another aged ram, unless I greatly alter my mind, or can find out some method of feeding them which will not destroy the animals, and which I have hitherto failed to accomplish." The conclusion which he adopted, in view of these liabilities, may be useful to agriculturists in America as well as in England. He says, "What I intend exhibiting in future will be shearlings only, as I believe they are not so easily injured by extra feeding as aged sheep, partly by being more active, and partly by having more time to put on their extra condition, by which their constitutions are not likely to be so much impaired."

At nearly every subsequent national exhibition, Mr. Webb carried off the best prizes for Southdowns. At Dundee, in 1843, the Highland Society paid him the compliment of having the likenesses of his sheep taken for its museum in Edinburgh. He only received two checks in these competitions after 1840, and these he rectified and overcame in an interesting way. The first took place at the great meeting at Exeter, in 1850, and the second at Chelmsford, in 1856. On both of these occasions he was convinced that the judges had not done justice to the qualities of his animals, and he resolved to submit their judgment to a court of errors, or to the decision of a subsequent meeting of the society. So, in 1851, he presented the unsuccessful candidate at Exeter to the meeting at Windsor, and took the first prize for it. This fully reversed the Exeter verdict. He resorted to the same tribunal to set him right in regard to his apparent defeat at Chelmsford, in 1856. Next year he presented the ram beaten there to the Salisbury meeting, and another jury gave the animal the highest meed of merit.

It was at the zenith of his fame as a sheep-breeder that Mr. Webb "assisted," as the French say, at the Universal Exposition at Paris, in 1855. Here his beautiful animals excited the liveliest admiration. The Emperor came himself to examine them, and expressed himself highly pleased at their splendid qualities. It was on this occasion that Mr. Webb presented to the Emperor his prize ram, for which, probably, he had refused the largest sum ever offered for a single animal of the same race, or 500 guineas ($2,500). The Emperor accepted the noble present, fully appreciating the spirit in which it was offered, and some time afterwards sent the generous breeder a magnificent candelabra, of solid silver, representing a grand, old English oak, with a group of horses shading themselves under its branches. This splendid token of the Emperor's regard is only one of the numerous trophies and souvenirs that embellish the farmer's home at Babraham, and which his children and remoter posterity will treasure as precious heirlooms.

If Mr. Webb did not originate, he developed a system of usefulness into a permanent and most valuable institution, which, perhaps, will be the most novel to American stock-raisers. Having, by a long course of scientific observations and experiments, *fixed* the qualities he desired to give his Southdowns; having brought them to the highest perfection, he now adopted a system which would most widely and cheaply diffuse the race thus cultivated all over the civilized world. He instituted an annual ram-letting, which took place in the month of July. This occasion constituted an important event to the great agricultural world. A few Americans have been present and witnessed the proceedings of these memorable days, and they know the interest attaching to them better than can be inferred from any description. M. De La Trehonnais, in the "Revue Agricole de l'Angleterre," thus sketches some of the incidents and aspects of the occasion:—

"It is a proceeding regarded in England as a public event, and all the journals give an account of it with exact care, assembling from every county and even from foreign countries. The sale begins about two o'clock. A circle in formed with ropes in a small field near the mansion, where the rams are introduced, and an auctioneer announces the biddings, which are frequently very spirited. The rams to be let are exposed around the field from the first of the morning, and a ticket at the head of each pen indicates the weight of the fleece of the animal it contains. Every one takes his notes, chooses the animal he approves of, and can demand the last bidding when he pleases. The evening after the letting, the numerous company assembles under a rustic shed, ornamented with leaves and agricultural devices. There tables are laid, around which are placed two or three hundred guests, and then commences one of those antique repasts described by Homer or Rabelais. The tables groan under the enormous pieces of beef, gigantic hams, etc., which have almost disappeared before the commencement of the sale. From eight in the morning until two in the afternoon, tables laid out in the dining-room and hall are furnished, only to be refurnished immediately, the end being equal to the beginning."

This description refers to the thirty-second letting. Mr. Webb's flock then consisted of seven hundred breeding ewes, a proportionate number of lambs, and about four hundred rams of different ages. It was from these rams that the animals were selected which were sent into every country in the civilized world. The average price of their lettings was nearly £24 each, although some of the rams brought the sum of £180, or nearly *nine hundred dollars*! What would some of the old-fashioned farmers of New England, of forty years ago, think of paying nearly a thousand dollars for the rent of a ram for a single year, or even one-tenth of that sum? But this rentage was not a fancy price. The farmer who paid it got back his money many times over in the course of a few years. From this infusion of the Babraham blood into his flock, he realised an augmented production of mutton and wool annually per acre which he could count definitely by pounds. The verdict of his balance-sheet proved the profit of the investment. It would be impossible to measure the benefit which the whole world reaped from Mr. Webb's labors in this department of usefulness. An eminent authority has stated that "it would be difficult, if not impossible, to find a Southdown flock of any reputation, in any country in the world, not closely allied with the Babraham flock. " It is a fact that illustrates the skill and care, as well as demonstrates the value of his system of improvement, that, after thirty-seven years as a breeder, the tribes he founded maintained to the last those distinguishing qualities which gave them such pre-eminence over all other sheep bearing the general name of the Sussex race. So valuable and distinctive were these qualities regarded by the best judges in the country, that the twelfth ram-letting, which took place at the time of the Cambridge Show, brought together 2,000 visitors, constituting, perhaps, the most distinguished assembly of agriculturists ever convened. On this occasion the Duke of Richmond, an hereditary and eminent breeder of Southdowns in their native county, bid a hundred guineas for a ram lamb, which Mr. Webb himself bought in.

Having attained to such eminence as a sheep-breeder, Mr. Webb entered upon another sphere of improvement, in which he won almost equal distinction. In 1837, he laid the foundation of the Babraham Herd of Shorthorn cattle, made up of six different tribes, purchased from the most valuable and celebrated branches of the race bearing that name. An incident attaching to one of these purchases may illustrate the nice care and cultivated skill which Mr. Webb exercised in the treatment of choice animals. He bought out of Lord Spencer's herd the celebrated cow, "Dodona." That eminent breeder, it appears, had given her up as irretrievably sterile, and he parted with her solely on that account. Mr. Webb, however, took her to Babraham, and, as a result of the more intelligent treatment he bestowed upon her, she produced successively four calves, which thus formed one of the most valuable families of the Babraham herd. When I visited the scene of his life and labors, all his sheep and cattle had been sold. But two or three animals bought by an Australian gentleman were still in the keeping of Mr. Webb's son, awaiting arrangements for their transportation. One of these, a beautiful heifer of fourteen months, was purchased at the winding-up sale, for 225 guineas. It was called the "Drawing-room Rose," from this circumstance, as I afterwards learned. When it was first dropped by the dam, Mr. Webb was confined to the house by indisposition. But he had such a desire to see this new accession to his bovine family, that he directed it to be brought into the drawing-room for that purpose. Hence

it received a more elegant and domestic appellation than the variegated nomenclature of high-blooded animals often allows.

When the last volume of the "English Herd-Book" was about to be published, Mr. Webb sent for insertion a list of sixty-one cows, with their products. He generally kept from twenty to thirty bulls in his stalls.

Nor were his labors confined even to the two great spheres of enterprise with which his name has been intimately and honorably associated. If it was the great aim of his intelligent activities to produce stock which should yield the most meat to the acre, he also gave great attention to the augmented production of the land itself. He was the principal originator and promoter of the great Agricultural Hall, in London, for the exhibition of the fat stock for the Smithfield Show. This may be called the Crystal Palace of the animal world. It is the grandest structure ever erected for the exhibition of cattle, sheep, swine, poultry, etc. I will essay no description of it here, but it will carry through long generations the name and memory of Jonas Webb of Babraham. He was chairman of the company that built the superb edifice; also president of the Nitro-phosphate or Blood-manure Company, a fertilizer in which he had the greatest confidence, and which he used in great quantities upon the large farm he cultivated, containing over 2,000 acres.

At the age of nearly sixty-six, Mr. Webb found that his health would no longer stand the strain of the toil, care, and anxiety requisite to keep up the Babraham flock to the high standard of perfection which it had attained. So, after nearly forty years of devotion to this great occupation of his life, he concluded to retire from it altogether, dispersing his sheep and cattle as widely as purchasers might be found. This breaking-up took place at Babraham on the 10th of July, 1862. Then and there the long series of annual re-unions terminated for ever. The occasion had a mournful interest to many who had attended those meetings from year to year. It seemed like the voluntary and unexpected abdi-

cation of an Alexander, still able to add to his conquests and trophies. All present felt this; and several tried to express it at the old table now spread for the last time for such guests. But his inherent and invincible modesty waived aside or intercepted the compliments that came from so many lips. With a kind of ingenious delicacy, which one of the finest of human sentiments could only inspire, he contrived to divert attention or reference to himself and his life's labors. But he could not make the company forget them, even if he gently checked allusion to them.

The company on this interesting occasion was very large, about 1,000 persons having sat down to the collation. Not only were the principal nobility and gentry of Great Britain interested in agricultural pursuits present in large number, but the representatives of nearly every other country in Christendom. Several gentlemen from the United States were among the purchasers. The total number of sheep sold was 969, which fetched under the hammer the great aggregate of £10,926, or more than $54,000. The most splendid ram in the flock went to the United States, being knocked down to Mr. J. C. Taylor, of Holmdale, New Jersey; who is doing so much to Americanise the Southdowns. Others went to the Canadas, Australia, South America, and to nearly every country in continental Europe.

Thus was formed, and thus was dispersed the famous Babraham flock. And such were the labors of Jonas Webb for the material well-being of mankind. These alone, detached from those qualities and characteristics which make up and reflect a higher nature, entitle his name to a wide and lasting memory among men. And these labors and successes are they that those who have read of them in different countries know him by. These comprise and present the character they honor with respect. What he was in the temper and disposition of his inner life, in daily walk and conversation, in the even and gentle amenities of Christian humility, in sudden trials of his faith and patience; what he was as a husband, father, friend and neighbor, to the poor, to the afflicted in mind, body or estate,—all this will remain unwritten, but not unremembered by those who breathed and moved within that disk of light which his life shed around him.

Few men have lived in whom so many personal and moral qualities combined to command respect, esteem, and even admiration. In stature, countenance, expression, and deportment, he was a noble specimen of fully developed English manhood. To this first, external aspect, his kindly and generous dispositions, his genial manners, his delicate but dignified modesty, his large intelligence and large-heartedness, gave the additional and crowning characteristic of a Christian gentleman. Many Americans have visited Babraham, and enjoyed the hospitalities which such a host could only give and grace. They will remember the paintings hung around the walls of that drawing-room, in which his commanding form, in the strength and beauty of meridian life, towers up in the rural landscape, surrounded by cattle and sheep bearing the impress of his skill and care. A little incident occurred a few years ago, which may illustrate this personal aspect better than any simile of description. On the occasion of one of the great Agricultural Expositions in Paris, a deputation or a company of gentlemen went over to represent the Agricultural Society of England. Mr. Webb was one of the number; and some French nobleman who had known him personally, as well as by reputation, was very desirous of making him a guest while in Paris. To be sure of this pleasure, he sent a special courier all the way to Folkestone, charged with a letter which he was himself to put into the hands of Mr. Webb, before the steamer left the dock. "But how am I to know the gentleman?" asked the courier; "I never saw him in my life." "*N'importe*," was the reply. "Put the letter in the hand of the noblest-looking man on board, and you will be sure to be right." The courier followed the direction; and, stationing himself near the gangway, he took his master's measure of every passenger as he en-

tered. He could not be mistaken. As soon as the plank was withdrawn, he approached Mr. Webb, hat in hand, and, with a deferential word of recognition, done in the best grace of French politeness, handed him the letter. One of the deputation, noticing the incident, and wondering how the man knew whom he was addressing without previous inquiry, questioned him afterwards on the subject, and learned from him the ground on which he proceeded. The photographic likeness presented in connection with this notice was taken shortly before his decease, at the age of nearly sixty-six, and when his health was greatly impaired.

Few men ever carried out so fully the injunction, not to let the left hand know what the right hand did, in the quiet and steady outflow of good will and good works, as Mr. Webb. Even those nearest and dearest to him never knew what that right hand did as a help in time of need, what that large heart felt in time of others' affliction, what those lips said to the sorrowing, in tearful moments of grief, until they had been stilled for ever on earth. Then it came out, act by act, word by word, thought by thought, from those who held the remembrances in their souls as precious souvenirs of a good man's life. So earnest was his desire to do these things in secret, that his own family heard of them only by accident, and from those whom he so greatly helped with his kindness and generosity. And when known by his wife and children, in this way, they were put under the bann of secrecy. This it is that makes it so difficult to delineate the home and heaven side of his character. Those nearest to him, who breathed in the blessing of its daily odor, so revere his repeated and earnest wish not to have his good works talked of in public, that, even now he is dead and gone, they hold it as a sacred obligation to his memory not to give up these treasured secrets of his life. Thus, in giving a partial *coup d'œil* of that aspect of his character which fronted homeward and heavenward, one can only glean, here and there, glimpses of different traits, in acts, incidents, and anec-

dotes remembered by neighbors and friends near and remote. Were it not that his children are withheld, by this delicate veneration, from giving to the public facts known to them alone, the moral beauty and brightness of his life would shine out upon the outside world with warmer rays and larger *rayons*. I hope that a single passage from a letter written by one of them to a friend, even under the injunction of confidence, may be given here, without rending the veil which they hold so sacred. In referring to this disposition and habit of her venerated father, she says—

"Often have I been so blessed as to be caused to shed tears of joy and pride at hearing proofs of his tenderness, kindness, and generosity related by the recipients of some token of his nobleness, but of which we never should have heard from himself."

A little incident may illustrate this trait of his disposition. In 1862, a "Loan Court" was held in London, at which there was a most magnificent display of jewels and plate of all kinds, contributed by their owners to be exhibited for the gratification of the public. A friend, who held him in the highest veneration, returning from this brilliant show, expressed regret that Mr. Webb had not furnished one of the stands, by sending the splendid silver candelabra presented to him by the French Emperor, with the many silver cups and medals he had won. Mr. Webb replied, that the mercies God had blessed him with, and the successes He had awarded to him, might have been sent to teach him humility, and not given to parade before the world.

It is one of the most striking proofs of his great and pure-heartedness, that, notwithstanding nearly forty consecutive years of vigorous and successful competition with the leading agriculturists of Great Britain and other countries, none of the victories he won over them, or the eminence he attained, ever made him an enemy. When we consider the eager ambitions and excited sensibilities that enter into these competitions, this fact in itself shows what manner of man he was in his disposition and deportment. Referring to this aspect of his character, the French writer already cited, M. De La Trehonnais, says of him, while still living—

"There exists no person who has gained the esteem and goodwill of his contemporaries to a higher degree than Mr. Webb. His probity, his scrupulous good faith, his generosity, and the affable equality of his character, have gained for him the respect and affection of every one. Since I have had the honor of knowing him, which is already many years, I have never known of his having a single enemy; and in my constant intercourse with the agricultural classes of England, I have never heard of a single malevolent insinuation respecting him. When we consider how much those who raise themselves in the world above others, are made the butt for the attacks of envy in proportion with their elevation, we may conclude that there are in the character of this wealthy man very solid virtues, well fixed principles, transcendant [sic] merit, to have passed through his long career of success and triumphs without having drawn upon himself the ill-will of a single enemy, or the calumnious shaft of envy."

Nor were these negative virtues, ending where they begun, or enabling him to go through a long life of energetic activities without an enemy. He not only lived at peace with all men, but did his utmost to make them live at peace with each other. Says one who knew him intimately—"I never heard him express a sentiment savoring of enmity to any person, nor could he bear to see it entertained by any one towards another. Even if he heard of an ill-feeling existing between persons, he would, if possible, effect a reconciliation; and his own bright example, and hearty, kind, genial manners always warmed all hearts towards himself. Notwithstanding the numerous calls upon his time, made by public and private business, he did not lose his sweet cheerfulness of temper, and was ever ready in his most busy moments to aid others, if he saw a possibility of so doing." Energy, gentleness, conscientiousness and courtesy were seldom, if ever, blended in such suave accord as in him. These virtues came out, each in its distinctive lustre, under the trials and vexations which try human nature most severely. All who knew him marvelled that he was able to maintain such sweetness and evenness of temper under provocations and difficulties which would have greatly annoyed most men. What he was in these outer circles of his influence, he was, to all the centralization of his virtues, in the heart of his family. Here, indeed, the best graces of his character had their full play and beauty. He was the centre and soul of one of the happiest of earthly homes, attracting to him the affections of every member of the hearth circle that moved in the sleepless light of his life. Here he did not rule, but led by love. It alone dictated, and it alone obeyed. It inspired its like in domestic discipline. Spontaneous reverence for such a father's wish and will superseded the unpleasant necessity of more active parental constraint. To bring a shade of sadness to that venerated face, or a speechless reproach to that benignant eye, was a greater punishment to a temporarily wayward child than any corporal correction could have inflicted.

No one of the hundreds that were present at the sale and dispersion of the Babraham flock could have thought that the remaining days of the great and good man were to be so few on earth. He was then about sixty-five years of age, of stately, unbending form and face radiant and genial with the florid flush of that Indian Summer which so many Englishmen wear late in those autumnal years that bend and pale American forms and faces to "the sere and yellow leaf" of life. But the sequel proved that he did not abdicate his position too early. In a little more than a year from this event, his spirit was raised to higher fellowships and folded with those of the pure and blest of bygone ages. The incidents and coincidents of the last, great moments of his being here, were remarkable and affecting. Neither he nor his wife died at the home they had made so happy with the beauty and savor of their virtues. Under another and distant roof they both laid themselves down to die. The husband's hand was linked in his wife's, up to within a few short steps of the river's brink, when, touched with the cold spray of the dark waters, it fell from its hold and was superseded by the strong arm of the angel of the covenant, sent to bear her fast across the flood. In life they were united to a oneness seldom witnessed on earth; in death they were not separated except by the thinnest partition. Though her spirit was taken up first to the great and holy communion above, the "ministering angel of God's love let her body remain with him as a pledge until his own spirit was called to join hers in the joint mansion of their eternal rest. On the very day that her body was carried to its long home, his own unloosed, to its upward flight, the soul that had made it shine for half a century like a temple erected to the Divine Glory. The years allotted to him on earth were even to a day. Just sixty-six were measured off to him, and then "the wheel ceased to turn at the cistern," and he died on his birthday. An affecting coincidence also marked the departure of his beloved wife. She left on the birthday of her eldest son, who had intended to make the anniversary the dating-day of domestic happiness, by choosing it for his marriage.

A few facts will suffice for the history of the closing scene. About the middle of October, 1862, Mrs. Webb, whose health seemed failing, went to visit her brother, Henry Marshall, Esq., residing in Cambridge. Here she suddenly became much worse, and the prospect of her recovery more and more doubtful. Mr. Webb was with her immediately on the first unfavorable turn of her illness, together with other members of the family. When he realised her danger, and the hope of her surviving broke down within him, his physical constitution succumbed under the impending blow, and two days before her death, he was prostrated by a nervous fever, from which he never rallied, but died on the 10th of November. Although the great visitation was too heavy for his flesh and blood to bear, his spirit was strengthened to drink this last cup of earthly trial with beautiful serenity and submission. It was strong enough to make his quivering lips to say, in distinct and audible utterance, and his closing eyes to pledge the truth and depth of the sentiment, "Thy will be done!" One who stood over him in these last moments says, that, when assured of his own danger, his countenance only seemed to take on a light of greater happiness. He was conscious up to within a few minutes of his death, and, though unable to speak articulately, responded by expressions of his countenance to the words and looks of affection addressed to him by the dear ones surrounding his bed. One of them read to him a favorite hymn, beginning with "Cling to the Comforter!" When she ceased, he signed to her to repeat it; and, while the words were still on her lips, the Comforter came at his call, and bore his waiting spirit away to the heavenly companionship for which it longed. As it left the stilled temple of its earthly habitation, it shed upon the delicately-carved lines of its marble door and closed windows a sweet gleam of the morning twilight of its own happy immortality.

A long funeral cortege attended the remains of the deceased from Cambridge to their last resting place in the little village churchyard of Babraham. Beside friends from neighboring villages, the First Cambridgeshire Mounted Rifle Corps joined the procession, together with a large number of the county police force. His body was laid down to its last, long rest beside that of his wife, who preceded him to the tomb only by a few days. Though Stratford-upon-Avon, and Dryburgh Abbey may attract more American travellers to their shrines, I am sure many of them, with due perception of moral worth, will visit Babraham, and hold it in reverent estimation as the home of one of the world's best worthies, who left on it a biograph which shall have a place among the human-life-scapes which the Saviour of mankind shall hang up in the inner temple of His Father's glory, as the most precious tokens and trophies of the earth, on which He shared the tearful experiences of humanity, and bore

back to His throne all the touching memories of its weaknesses, griefs, and sorrows.

A movement is now on foot to erect a suitable monument to his memory. It may indicate the public estimation in which his life and labors are held that, already, about £10,000 have been subscribed towards this testimonial to his worth. The monument, doubtless, will be placed in the great Agricultural Hall, which he did so much to found. His name will wear down to coming generations the crystal roofage of that magnificent edifice as a fitting crown of honor.

CHAPTER VIII.

THRESHING MACHINE—FLOWER SHOW—THE HOLLYHOCK AND ITS SUGGESTIONS—THE LAW OF CO-OPERATIVE ACTIVITIES IN VEG-ETABLE, ANIMAL, MENTAL, AND MORAL LIFE.

"In all places, then, and in all seasons,
Flowers expand their light and soul-like wings,
Teaching us, by most persuasive reasons,
How akin they are to human things."—
LONGFELLOW.

My stay at Babraham was short. It was like a visit to the grave of one of those English worthies whose lives and labors are so well known and appreciated in America. All the external features of the establishment were there unchanged. The large and substantial mansion, with its hall and parlor walls hung with the mementoes of the genius and success that had made it so celebrated; the barns and housings for the great herds and flocks which had been dispersed over the world; the very pens still standing in which they had been folded in for the auctioneer's hammer; all these arrangements and aspects remained as they were when Jonas Webb left his home to return no more. But all those beautiful and happy families of animal life, which he reared to such

perfection, were scattered on the wings of wind and steam to the uttermost and most opposite parts of the earth.

The eldest son, Mr. Samuel Webb, who supervises part of the farm occupied by his father, and also carries on one of his own in a neighboring parish, was very cordial and courteous, and drove me to his establishment near Chesterford. Here a steam threshing machine was at work, doing prodigious execution on different kinds of grain. The engine had climbed, *a proprii motu*, a long ascent; had made its way partly through ploughed land to the rear of the barn, and was rattlingly busy in a fog of dust, doing the labor of a hundred flails. Ricks of wheat and beans, each as large as a comfortable cottage, disappeared in quick succession through the fingers of the chattering, iron-ribbed giant, and came out in thick and rapid streams of yellow grain. Swine seemed to be the speciality to which this son of Mr. Webb is giving some of that attention which his father gave to sheep. There were between 200 and 300 in the barn-yards and pens, of different ages and breeds, all looking in excellent condition.

From Chesterford I went on to Cambridge, where I remained for the most part of two days, on account of a heavy fall of rain, which kept me within doors nearly all the time. I went out, however, for an hour or so to see a Flower Show in the Town Hall. The varieties and specimens made a beautiful, but not very extensive array. There was one flower that not only attracted especial admiration, but invited a pleasant train of thoughts to my own mind. It was one of those old favorites to which the common people of all countries, who speak our mother tongue, love to give an inalienable English name—*The Hollyhock*. It is one of the flowers of the people, which the pedantic Latinists have left untouched in homely Saxon, because the people would have none of their long-winded and heartless appellations. Having dwelt briefly upon the honor that Divine Providence confers upon human genius and labor, in letting them impress their finger-marks so distinctly upon the features and functions of the earth, and upon the forms of animal life, it may be a profitable recurrence to the same line of thought to notice what that same genius and labor have wrought upon the structure and face of this familiar flower. What was it at first? What is it now in the rural gardens of New England? A shallow, bell-mouthed cup, in most cases purely white, and hung to a tall, coarse stalk, like the yellow jets of a mullein. That is its natural and distinctive characteristic in all countries; at least where it is best known and most common. What is it here, bearing the fingerprints of man's mind and taste upon it? Its white and thin-sided cup is brim full and running over with flowery exuberance of leaf and tint infinitely variegated. Here it is as solid, as globe-faced, and nearly as large as the dahlia. Place it side by side with the old, single-leafed hollyhock, in a New England farmer's garden, and his wife would not be able to trace any family relationship between them, even through the spectacles with which she reads the Bible. But the dahlia itself—what was that in its first estate, in the country in which it was first found in its aboriginal structure and complexion? As plain and unpretending as the hollyhock; as thinly dressed as the short-kirtled daisy in a Connecticut meadow. It is wonderful, and passing wonder, how teachable and quick of perception and prehension is Nature in the studio of Art. She, the oldest of painters, that hung the earth, sea, and sky of the antediluvian world with landscapes, waterscapes, and cloudscapes manifold and beautiful, when as yet the human hand had never lifted a pencil to imitate her skill; she, with the colors wherewith she dyed the fleecy clouds that spread their purple drapery over the first sunset, and in which she dipped the first rainbow hung in heaven, and the first rose that breathed and blushed on earth; she that has embellished every day, since the Sun first opened its eye upon the world, with a new gallery of paintings for every square mile of land and sea, and new dissolving views for every hour—she, with all these artistic antecedents,

tastes, and faculties, comes modestly into the conservatory of the floriculturist, and takes lessons of him in shaping and tinting plants and flowers which the Great Master said were "all very good" on the sixth-day morning of the creation! This is marvellous, showing a prerogative in human genius almost divine, and worthy of reverent and grateful admiration. How wide-reaching and multigerent is this prerogative! In how many spheres of action it works simultaneously in these latter days! See how it manipulates the brute forces of Nature! See how it saddles the winds, and bridles and spurs the lightning! See how it harnesses steam to the plough, the flood to the spindle, the quick cross currents of electricity to the newsman's phaeton! Then ascend to higher reaches of its faculty. In the hands of a Bakewell or Webb, it gives a new and creative shaping to multitudinous generations of animal life. Nature yields to its suggestion and leading, and co-works, with all her best and busiest activities, to realise the human ideal; to put muscle there, to straighten that vertebra, to parallel more perfectly those dorsal and ventral lines, to lengthen or shorten those bones; to flesh the leg only to such a joint, and wool or unwool it below; to horn or unhorn the head, to blacken or blanch the face, to put on the whole body a new dress and make it and its remote posterity wear this new form and costume for evermore. All this shows how kindly and how proudly Nature takes Art into partnership with her, in these new structures of beauty and perfection; both teaching and taught, and wooing man to work with her, and walk with her, and talk with her within the domain of creative energies; to make the cattle and sheep of ten thousand hills and valleys thank the Lord, out of the grateful speech of their large, lustrous eyes, for better forms and features, and faculties of comfort than their early predecessors were born to.

Equally wonderful, perhaps more beautiful, is the joint work of Nature and Art on the sweet life and glory of flowers. However many they were, and what they were, that breathed upon the

first Spring or Summer day of time, each was a half-sealed gift of God to man, to be opened by his hand when his mind should open to a new sense of beauty and perfection. Flowers, each with a genealogy reaching unbroken through the Flood back to the overhanging blossoms of Eden, have come down to us, as it were, only in their travelling costume, with their best dresses packed away in stamen, or petal, or private seedcase, to be brought out at the end of fifty centuries at the touch of human genius. Those of which Solomon sang in his time, and which exceeded his glory in their every-day array, even "the hyssop by the wall," never showed, on the gala-days of his Egyptian bride, the hidden charms which he, in his wisdom, knew not how to unlock. Flowers innumerable are now, like illuminated capitals of Nature's alphabet, flecking, with their sheen-dots, prairie, steppe, mountain and meadow, the earth around, which, perhaps, will only give their best beauties to the world in a distant age. As the light of the latest-created and remotest stars has not yet completed its downward journey to the eye of man, so to his sight have not these sweet-breathing constellations of the field yet made the full revelation of their treasured hues and forms. Not one in a hundred of them all has done this up to the present moment. When one in ten of those that bless us with their life and being shall put on all its reserved beauty, then, indeed, the stars above and the stars below will stud the firmaments in which they shine with equal glory, and blend both in one great heaven-scape for the eye and heart of man. One by one, in its turn, the key of human genius shall unlock the hidden wardrobe of the commonest flowers, and deck them out in the court dress reserved, for five thousand years, to be worn in the brighter, afternoon centuries of the world. The Mistress of the Robes is a high dignitary in the Household of Royalty, and has her place near to the person of the Queen. But the Floriculturist, of educated perception and taste, is the master of a higher state robe, and holds the key of embroidered vestments, cosmetics, tintings, artistries, hair-jewels, head-dresses, brooches, and bracelets, which no empress ever wore since human crowns were made; which Nature herself could not show on all the bygone birthdays of her being.

This is marvellous. It is an honor to man, put upon him from above, as one of the gratuitous dignities of his being. "An undevout astronomer is mad," said one who had opened his mind to a broad grasp of the wonders which this upper heaven holds in its bosom. The floriculturist is an astronomer, with Newton's telescope reversed; and if its revelations do not stir up holy thoughts in his soul, he is blind as well as mad. No glass, no geometry that Newton ever lifted at the still star-worlds above, could do more than *reveal*. At the farthest stretch of their faculty, they could only bring to light the life and immortality of those orbs which the human eye had never seen before. They could not tint nor add a ray to one of them all. They never could bring down to the reach of man's unaided vision a single star that Noah could not see through the deck-lights of the ark. It was a gift and a glory that well rewarded the science and genius of Newton and Herschel, of Adams and Le Verrier, that they could ladder these mighty perpendicular distances and climb the rounds to such heights and sweeps of observation, and count, measure, and name orbs and orbits before unknown, and chart the paths of their rotations and weigh them, as in scales, while in motion. But this *ge*-astronomer, whose observatory is his conservatory, whose telescope and fluxions are his trowel and watering-pot, not only brings to light the hidden life of a thousand earth-stars, but changes their forms, colors their rays, half creates and transforms, until each differs as much from its original structure and tinting as the planet Jupiter would differ from its familiar countenance if Adams or Le Verrier could make it wear the florid face of Mars. This man,—and it is to be hoped he carries some devout and grateful thoughts to his work—sets Nature new lessons daily in artistry, and she works out the new ideals of his taste

to their joint and equal admiration. He has got up a new pattern for the fern. She lets him guide her hand in the delicate operation, and she crimps, fringes, shades or shapes its leaflets to his will, even to a thousand varieties. He moistens her fingers with the fluids she uses on her easel, and puts them to the rootlets of the rose, and they transpose its hues, or fringe it or tinge it with a new glory. He goes into the fen or forest, or climbs the jutting crags of lava-mailed mountains, and brings back to his fold one of Nature's foundlings,—a little, pale-faced orphan, crouching, pinched and starved, in a ragged hood of dirty muslin; and he puts it under the fostering of those maternal fingers, guided by his own. Soon it feels the inspiration of a new life warming and swelling its shrivelled veins. Its paralysed petals unfold, one by one. The rim of its cup fills, leaf by leaf, to the brim. It becomes a thing most lovely and fair, and he introduces it, with pride, to the court beauties of his crystal palace.

The agriculturist is taken into this co-partnership of Nature in a higher domain of her activities, measured by the great utilities of human life. We have glanced at the joint-work in her animal kingdom. In the vegetable, it is equally wonderful. Nature contributes the raw material of these great and vital industries, then incites and works out human suggestions. Thus she trains and obeys the mind and hand of man, in this grand sphere of development. Their co-working and its result are just as perceptible in a common Irish potato as in the most gorgeous dahlia ever exhibited. Not one farmer in a thousand has ever read the history of that root of roots, in value to mankind; has ever conceived what a tasteless, contracted, water-soaked thing it was in its wild and original condition. Let them read a few chapters of the early history of New England, and they will see what it was two hundred and fifty years ago, when the strong-hearted men and women, whom Hooker led to the banks of the Connecticut, sought for it in the white woods of winter, scraping away the snow with their frosted fingers. The largest they found

just equalled the Malaga grape in size and resembled it in complexion. They called it the *ground-nut*, for it seemed akin to the nuts dropped by the oaks of different names. No flower that breathes on earth has been made to produce so many varieties of form, complexion, and name as this homely root. It would be an interesting and instructive enterprise, to array all the varieties of this queen of esculent vegetables which Europe and America could exhibit, face to face with all the varieties which the dahlia, geranium, pansy, or even the fern has produced, and then see which has been numerically the most prolific in diversification of forms and features. It should gratify a better motive than curiosity to trace back the history of other roots to their aboriginal condition. Types of the original stock may now be found, in waste places, in the wild turnip, wild carrot, parsnip, etc. "Line upon line, precept upon precept, here a little and there a little," it may be truly and gratefully said, these roots, internetted with the very life-fibres of human sustenance, have been brought to their present perfection and value. The great governments and peoples of the world should give admiring and grateful thought to this fact. Here nature co-works with the most common and inartistic of human industries, as they are generally held, with faculties as subtle and beautiful as those which she brings to bear upon the choicest flowers. The same is true of grains and grasses for man and beast. They come down to us from a kind of heathen parentage, receiving new forms and qualities from age to age. The wheats, which make the bread of all the continents, now exhibit varieties which no one has undertaken to enumerate. Fruits follow the same rule, and show the same joint-working of Nature and Art as in the realm of flowers.

The wheel within wheel, the circle within circle expand and ascend until the last circumferential line sweeps around all the world of created being, even taking in, upon the common radius, the highest and oldest of the angels. From the primrose peering from the hedge to the premier seraph wearing the coronet of his sublime companionship; from the lowest forms of vegetable existence to the loftiest reaches of moral nature this side of the Infinite, this everlasting law of co-working rules the ratio of progress and development. In all the concentric spheres strung on the radius measured by these extremes, there is the same co-acting of internal and external forces. And mind, of man or angel, guides and governs both. Not a flower that ever breathed on earth, not one that ever blushed in Eden, could open all its hidden treasures of beauty without the co-working of man's mind and taste. No animal that ever bowed its neck to his yoke, or gave him labor, milk or wool, could come to the full development of its latent vitalities and symmetries without the help of his thought and skill. The same law obtains in his own physical nature. Mind has made it what it is to-day, as compared with the wild features and habits of its aboriginal condition. Mind has worked for five thousand years upon its fellow-traveller through time, to fit it more and more fully for the companionship. It was delivered over to her charge naked, with its attributes and faculties as latent and dormant as those of the wild rose or dahlia. Through all the ages long, she has worked upon its development; educating its tastes; taming its appetites; refining its sensibilities; multiplying and softening its enjoyments; giving to every sense a new capacity and relish of delight; cultivating the ear for music, and ravishing it with the concord of sweet sounds; cultivating the eye to drink in the glorious beauty of the external world, then adding to natural sceneries ten thousand pictures of mountain, valley, river, man, angel, and scenes in human and heaven's history, painted by the thought-instructed hand; cultivating the palate to the most exquisite sensibilities, and exploring all the zones for luxuries to gratify them; cultivating the fine finger-nerves to such perception that they can feel the pulse of sleeping notes of music; cultivating the still finer organism that catches the subtle odors on the wing, and sends their separate or min-

gled breathings through every vein and muscle from head to foot.

The same law holds good in the development of mind. It has now reached such an altitude, and it shines with such lustre, that our imagination can hardly find the way down to the morning horizon of its life, and measure its scope and power in the dim twilight of its first hours in time. The simple fact of its first condition would now seem to most men as exaggerated fancies, if given in the simplest forms of truthful statement. With all the mighty faculties to which it has come; with its capacity to count, name, measure and weigh stars that Adam, nor Moses, nor Solomon ever saw; with all the forces of nature it has subdued to the service of man, it cannot tell what simplest facts of the creation had to be ascertained by its first, feeble and confused reasonings. No one of to-day can say how low down in the scale of intelligence the human mind began to exercise its untried faculties; what apposition and deduction of thoughts it required to individualise the commonest objects that met the eye; even to determine that the body it animated was not an immovable part of the earth itself; to obtain fixed notions of distance, of color, light, and heat; to learn the properties and uses of plants, herbs, and fruits; even to see the sun sink out of sight with the sure faith that it would rise again. It was gifted with no instinct, to decide these questions instantly and mechanically. They had all to pass through the varied processes of reason. The first bird that sang in Eden, built its first nest as perfectly as its last. But, thought by thought, the first human mind worked out conclusions which the dullest beast or bird reached instantly without reason. What wonderful co-working of internal and external influences was provided to keep thought in sleepless action; to open, one by one, the myriad petals of the mind! Nature, with all its shifting sceneries, filled every new scope of vision with objects that hourly set thought at play in a new line of reflection. Then, out of man's physical being came a thousand still small voices daily, whispering, Think! think! The first-born ne-

cessities, few and simple, cried, "*Think*! for we want bread, we want drink, we want shelter and raiment against the cold." The finer senses cried continually, "Give! give thought to this, to that." The Eye, the Ear, the Palate and every other organ that could receive and diffuse delight, worked the mental faculties by day and night, up to the last sunset of the antediluvian world; and all the intellectual result of this working Noah took with him into the ark, and gave to his sons to hand over to succeeding ages. Flowers that Eve stuck in the hair of the infant Abel are just now opening the last casket of their beauty to the favored children of our time. This, in itself, is a marvellous instance of the law we are noticing. But what is this to the processes of thought and observation through which the mind of man has reached its present expansion; through which it has developed all these sciences, arts, industries and tastes, the literature and the intellectual life of these bright days of humanity! The figure is weak, and every figure would be weak when applied to the ratio or the result of this progression; but, at what future age of time, or of the existence beyond time, will the mind, that has thus wrought on earth, open its last petal, put forth no new breathing, unfold no new beauty under the eye of the Infinite, who breathed it, as an immortal atom of His own essence, into the being of man?

Follow the radius up into the next concentric circle, and we see this law working to finer and sublimer issues in man's moral nature. We have glanced at what the mind has done for and through his physical faculties and being; how that being has re-acted upon the mind, and kept all its capacities at work in procuring new delight to the eye, ear, palate, and all the senses that yearned for enjoyment. We have noticed how the inside and outside world acted upon his reasoning powers in the dawn of creation; how slowly they mastered the simplest facts and phenomena of life in and around them, how slowly they expanded, through the intervening centuries, to their present development. The mind is the central personage in the trinity of man's being; linking the mortal and immortal to its life and action; vitalising the body with intelligence, until every vein, muscle, and nerve, and function thrills and moves to the impulse of thought; vitalising the soul with the vigorous activities of reason, giving hands as well as wings to its hopes, faiths, loves, and aspirations; giving a faculty of speech, action, and influence to each, and play to all the tempers and tendencies of its moral nature. Thus all the influences that the mind could inhale from the material world through man's physical being, and all it could draw out of the depths of Divine revelation, were the dew and the light which it was its mission to bring to the fostering, growth, and glory of the human soul. These were man's means wherewith to shape it for its great destiny; these he was to bring to its training and expansion; with these he was to co-work with the Infinite Father of Spirits to fit it for His presence and fellowship, just as he co-works with Nature in developing the latent life and faculties of the rose. What distillations of spiritual influence have dropped down out of heaven, through the ages, to help onward this joint work! What histories of human experience have come in the other direction to the same end!—fraught with the emotions of the human heart, from the first sin and sorrow of Adam to our own griefs, hopes, and joys; and all so many lessons for the discipline of this highborn nature with us!

And yet how slow and almost imperceptible has been the development of this nature! How gently and gradually the expanding influences, human and divine, have been let in upon its latent faculties! See with what delicate fostering the petals of love, faith, and hope were taught to open, little by little, their hidden life and beauty,—taking Moses' history of the process. First, one human being on the earth, surrounded with beasts and birds that could give him no intelligent companionship and no fellow-feeling. Then the beautiful being created to meet these awakening yearnings of his nature; then the first outflow and interchange of human love. The

narrative brings us to the next stage of the sentiment. Sin and sorrow afflict, but unite, both hearts in the saddest experience of humanity. They are driven out of the Eden of their first condition, but their very sufferings and fears re-Eden their mutual attachments in the very thorns of their troubles and sorrows. Then another being, of their own flesh, heir to their changed lot, and to these attachments, is added to their companionship. The first child's face that heaven or earth ever saw, opened its baby eyes on them and smiled in the light of their parental love. The history goes on. In process of time, there is a family of families, called a community, embracing hundreds of individuals connected by ties of blood so attenuated that they possess no binding influence. Common interests, affinities, and sentiments supply the place of family relationship, and make laws of amity and equity for them as a population. Next we have a community of communities, or a commonwealth of these individual populations, generally called a nation. Here is a lesson for the moral nature. Here are thousands and tens of thousands of men who never saw each others' faces. Will this expanded orb of humanity revolve around the same centre as the first family circle, or the first independent community? How can you give it cohesion and harmony? Extend the *radii* of family relationship and influence to its circumference in every direction. Throne the sovereign in a parent's chair, to execute a father's laws. He shall treat them as children, and they each other as brethren. Here is a grand programme for human society. Here is a vigorous discipline for the wayward will and temper of the human heart. How is a man to feel and act in these new conditions? How is he to regulate his hates and loves, his passions and appetites, to comply properly with these extended and complicated relationships?

About half way from Adam's day to ours, there came an utterance from Mount Sinai that anticipated and answered these questions once for all, and for one and all. In that august revelation

of the Divine Mind, every command of the Decalogue swung open upon the pivot of a *not*, except one; and that one referred to man's duty to man, and the promise attached to its fulfilment was only an earthly enjoyment. All the rest were restrictive; to curb this appetite, to bar that passion, to hedge this impulse, to check that disposition; in a word, to hold back the hand from open and positive transgression. Even the first, relating to His own Godhead and requirements, was but the first of the series of negatives, a pure and simple prohibition of idolatry. No reward of keeping this first great law, reaching beyond the boundary of a temporal condition, was promised at its giving out. With the headstrong passions, lusts, appetites, and tempers of flesh and blood bridled and bitted by these restrictions, and with no motives to obedience beyond the awards of a short life on earth, the human soul groped its way through twenty centuries after the Revelation of Sinai, feeling for the immortality which was not yet revealed to it, even "as through a glass darkly." Here and there, but thinly scattered through the ages, divinely illumined men caught, through the parting seams of the veil, a transient glimpse and ray of the life to come. Here and there, obscurely and hesitatingly, they refer to this vision of their faith. Here and there we seem to see a hope climbing up out of a good man's heart into the pathless mystery of a future existence, and bringing back the fragment of a leaf which it believes must have grown on one of the trees of life immortal. Moses, Job, David, and Isaiah give us utterances that savor of this belief; but they leave us in the dark in reference to its influence upon their lives. We cannot glean from these incidental expressions, whether it brought them any steady comfort, or sensibly affected their happiness.

Thus, for four thousand years, the soul of man dashed its wings against the prison-bars of time, peering into the night through the cold, relentless gratings for some fugitive ray of the existence of which it had such strong and sleepless presentiment. It is a mystery. It may seem irreverent to approach it even with a conjecture. Human reason should be humble and silent before it, and close its questioning lips. It may not, however, transcend its prerogative to say meekly, *perhaps*. Perhaps, then, for two-thirds of the duration that the sun has measured off to humanity, that life and immortality which the soul groped after were veiled from its vision, until all its mental and spiritual faculties had been trained and strengthened to the ability to grasp and appropriate the great fact when it should be revealed. Perhaps it required all the space of forty centuries to put forth feelers and fibres capable of clinging to the revelation with the steady hold of faith. Perhaps it was to prove, by long, decisive probation, what the unaided human mind could do in constructing its idealisms of immortality. Perhaps it was permitted to erect a scaffolding of conceptions on which to receive the great revelation at the highest possible level of thought and instinctive sentiment to which man could attain without supernatural light and help. If this last *perhaps* is preferable to the others, where was this scaffolding the highest? Over Confucius, or Socrates, or the Scandinavian seer, or Druid or Aztec priest? Was it highest at Athens, because there the great apostle to the Gentiles planted his feet upon it, and said, in the ears of the Grecian sophists, "Him whom ye ignorantly worship declare I unto you?" At that brilliant centre of pagan civilization it might have reached its loftiest altitude, measured by a purely intellectual standard; but morally, this scaffolding was on the same low level of human life and character all the world around. The immortalities erected by Egyptian or Grecian philosophy were no purer, in moral conception and attributes, than the mythological fantasies of the North American Indians. In them all, human nature was to have the old play of its passions and appetites; in some of them, a wider sweep and sway. There was not one in the whole set of Grecian deities half so moral and pure, in sentiment and conduct, as Socrates; nor were Jupiter and his subordinate celestials better than the

average kings and courts of Greece. Out of the hay, wood, and stubble of sheer fancy the human mind was left to raise these fantastic structures. They exercised and entertained the imagination, but brought no light nor strength to the soul; no superior nor additional motives to shape the conduct of life. But they did this, undoubtedly, with all their delusions; they developed the *thought* of immortality among the most benighted races of men. Their most perplexing unrealities kept the mind restless and almost eager for some supplementary manifestation; so that, when the Star of Bethlehem shone out in the sky of Palestine, there were men looking heavenward with expectant eyes at midnight. From that hour to this, and among pagan tribes of the lowest moral perception, the heralds of the Great Revelation have found the *thought* of another existence active though confused. They have found everywhere a platform already erected, like that on which Paul stood in the midst of Mars Hill, and on which they could stand and say to heathen communities, "Him whom ye ignorantly worship declare I unto you! That future life and immortality which your darkened eyes and hungry souls have been groping and hungering for, bring we to you, bright as the sun, in this great gospel of Divine Love." Had the Star of Bethlehem appeared a century earlier, it might not have met an upturned eye. If the Saviour of Mankind had come into the world in Solomon's day, not even a manger might have been found to cradle His first moments of human life; no Simeon waiting in the temple to greet the great salvation He brought to our race in His baby hands.

Here, then, commences, as it were, the central era of the soul's training in time. Here heaven opened upon it the full sunlight and sunwarmth of its glorious life and immortality. Here fell upon its opening faculties the dews and rays and spiritual influences which were to shape its being and destiny. Here commenced such co-working to this end as can find no measure nor simile in any other sphere of co-operative activities in the world below or above. Here the

trinity of man and the Trinity of the Godhead came into a co-action and fellowship overpassing the highest outside wonder of the universe. And all this co-working, fellowship, and partnership has been repeated in the experience of every individual soul that has been fitted for this great immortality. Here, too, this co-working is a law, not an incident; most marvellously, mightily, and minutely a law, as legislatively and executively as that which we have seen acting upon the development of the flower. Had not the great apostle, who was caught up into the third heavens and heard things unutterable, spoken of this law in such bold words, it would seem rash and irreverent in us to approach so near to its sublime revelation. Not ours but his they are; and it is bold enough in us to repeat them. He said it: that He, to whose name every knee should bow, and every tongue confess; to whom belonged and who should possess and rule all the kingdoms of the earth, "was made under the law," not of Moses, not of human nature only, but under this very law of CO-WORKING. Through this the world was to be regenerated and filled with His life and light. Through this a new creation was to be enfolded in the bosom of His glory, of grander dimensions and of diviner attributes than that over which the morning stars sang at the birth of time. Said this law to the individual soul, "Work out your salvation with fear and trembling, for it is God that worketh in you to will and to do of His own good pleasure." To will and to do. It is His own good will and pleasure that the soul shall be fitted and lifted up to its high destiny through this co-working. It was His power to raise it to that condition without man's participation or conscious acquiescence; but it was His will and pleasure to enact this law of salvation. Looking across the circumference of the individual soul, what says this law? "Go ye out into all the world and preach the gospel to every creature, and, lo, I am with you unto the end,"—not as an invisible companion, not merely with the still, small voice of the Comforter to cheer you in trial, weakness and privation; but with you as a *co-worker*, with the irresistible energies of the Spirit of Power. He might have done the whole work alone. He might have sent forth twelve, and twelve times twelve legions of angels, and given each a voice as loud as his who is to wake the dead, and bid them preach His gospel in the ears of every human being. He might have given a tongue to every breathing of the breeze, an articulate speech to every ray of light, and sent them out with their ceaseless voices on the great errand of His love. It was his power to do this. He did not do it, because it was His will and pleasure to put Himself under this law we have followed so far; to make men His co-workers in this new creation, and co-heirs with Him in all its joy and glory. So completely has He made this law His rule of action, that, for eighteen hundred years, we have not a single instance in which the life and immortality which He brought to light have been revealed to a human soul without the direct and active participation of a human instrumentality. So completely have His meekest servants on earth put themselves under this law, that not one of them dares to expect, hope, or pray that He will reveal Himself to a single benighted heathen mind without this human co-working.

Thus, begin where you will, in the flower of the field or the hyssop by the wall, and ascend from sphere to sphere, until there is no more space in things and beings created to draw another circumferential line, and you will see the action and the result of this great law of *Co-operative Activities*. When I first looked within the lids of that hollyhock, and was incited to read the rudimental lessons of the new leaves that man's art had added to its scant, original volume, I had no thought of finding so much matter printed on its pages. I have transcribed it here in the order of its paragraphs, hoping that some who read them may see in this life of flowers an interest they may have partially overlooked.

CHAPTER IX.

VISIT TO A THREE-THOUSAND-ACRE FARM—SAMUEL JONAS—HIS AGRICULTURAL OPERATIONS, THEIR EXTENT, SUCCESS, AND GENERAL ECONOMY.

The rain having ceased, I resumed my walk, in a southerly direction, to Chrishall Grange, the residence of Samuel Jonas, who may be called the largest farmer in England; not, perhaps, in extent of territory occupied, but in the productive capacity of the land cultivated, and in the values realised from it. It is about four miles east of Royston, bordering on the three counties of Cambridgeshire, Hertfordshire, and Essex, though lying mainly in the latter. It contains upwards of 3,000 acres, and nearly every one of them is arable, and under active cultivation. It consists of five farms, belonging to four different landlords; still they are so contiguous and coherent that they form substantially one great block. No one could be more deeply impressed with the magnitude of such an establishment, and of the operations it involves, than a New England farmer. Taking the average of our agriculturists, their holdings or occupations, to use an English term, will not exceed 100 acres each; and, including woodland, swamp, and mountain, not over half of this space can be cultivated. To the owner and tiller of such a farm, a visit to Mr. Jonas' occupation must be interesting and instructive. Here is a man who cultivates a space which thirty Connecticut farmers would feel themselves rich to own and occupy, with families making a population of full two hundred souls, supporting and filling a church and school-house. In the great West of America, where cattle are bred and fed somewhat after the manner of Russian steppes or Mexican ranches, such an occupation would not be unusual nor unexpected; but in the very heart of England, containing a space less than the state of Virginia, a tract of such extent and value in the hands of a single farmer is a fact which a New Englander must regard at first with no little sur-

prise. He will not wonder how one man can *rent* such a space, but how he can *till* it to advantage; how, even with the help of several intelligent and active sons, he can direct and supervise operations which fill the hands of thirty solid farmers of Massachusetts. Two specific circumstances enable him to perform this undertaking.

In the first place, agriculture in England is reduced to an exact and rigid science. To use a nautical phrase, it is all plain sailing. The course is charted even in the written contract with the landlord. The very term, "*course*," is adopted to designate the direction which the English farmer is to observe. Skilled hands are plenty and pressing to man the enterprise. With such a chart, and such a force, and such an open sea, it is as easy for him to sail the "Great Eastern" as a Thames schooner. The helm of the great ship plays as freely and faithfully to the motion of his will as the rudder of the small craft. Then the English farmer has a great advantage over the American in this circumstance: he can hire cheaply a grade of labor which is never brought to our market. Men of great skill and experience, who in America would conduct farms of their own, and could not be hired at any price, may be had here in abundance for foremen, at from twelve to sixteen shillings, or from three to four dollars a week, they boarding and lodging themselves. And the number of such men is constantly increasing, from two distinct causes. In the first place there is a large generation of agricultural laborers in England, now in the prime of manhood, who have just graduated, as it were, through all the scientific processes of agriculture developed in the last fifteen years. The ploughmen, cowmen, cartmen, and shepherds, even, have become familiar with the established routine; and every set of these hands can produce one or two active and intelligent laborers who will gladly and ably fill the post of under-foreman for a shilling or two a week of advanced wages. Then, by the constant absorption of small holdings into large farms, which is going on more rapidly from this increased facility of managing great occupations, a very considerable number of small farmers every year are falling into the labor market, being reduced to the necessity of either emigrating to cheaper lands beyond the sea, or of hiring themselves out at home as managers, foremen or common laborers on the estates thus enlarged by their little holdings. From these two sources of supply, the English tenant-farmer, beyond all question, is able to cultivate a larger space, and conduct more extensive operations than any other agriculturist in the world, at least by free labor.

The first peculiarity of this large occupation I noticed, was the extent of the fields into which it was divided. I had never seen any so large before in England. There were only three of the whole estate under 60, and some contained more than 400 acres each, giving the whole an aspect of amplitude like that of a rolling prairie farm in Illinois. Not one of the little, irregular morsels of land half swallowed by its broad-bottomed hedging, which one sees so frequently in an English landscape, could be found on this great holding. The white thorn fences were new, trim, and straight, occupying as little space as possible. The five amalgamated farms are light turnip soil, with the exception of about 200 acres, which are well drained. The whole surface resembles that of a heavy ground swell of the sea; nearly all the fields declining gently in different directions. The view from the rounded crest of the highest wave was exceedingly picturesque and beautiful, presenting a vista of plenty which Ceres of classic mythology never saw; for never, in ancient Greece, Italy, or Egypt, were the crops of vegetation so diversified and contrasting with each other as are interspersed over an English farm of the present day.

It is doubtful if 3,000 acres of land, lying in one solid block, could be found in England better adapted for testing and rewarding the most scientific and expensive processes of agriculture, than this great occupation of Mr. Jonas. Certainly, no equal space could present a less quantity of waste land, or occupy less in hedges or fences. And it is equally certain that no estate of equal size is more highly cultivated, or yields a greater amount of production per acre. Its occupant, also, is what may be called an hereditary farmer. His father and his remote ancestors were farmers, and he, as in the case of the late Mr. Webb, has attained to his present position as an agriculturist by practical farming.

Mr. Jonas cultivates his land on the "Four-course system." This very term indicates the degree to which English agriculture has been reduced to a precise and rigid science. It means here, that the whole arable extent of his estate is divided equally between four great crops; or, wheat, 750 acres; barley and oats, 750; seeds and pulse, 750; and roots, 750. Now, an American farmer, in order to form an approximate idea of the amount of labor given to the growth of these crops, must remember that all these great fields of wheat, oats, barley, turnips, beans, and peas, containing in all over 2,000 acres, are hoed by hand once or twice. His cereals are all drilled in at seven inches apart, turnips at seventeen. The latter are horse-hoed three or four times; and as they are drilled on the flat, or without ridging the surface of the ground, they are crossed with a horse-hoe with eight V shaped blades. This operation leaves the plants in bunches, which are singled out by a troop of children. One hand-hoeing and two or three more horse-hoeings finish the labor given to their cultivation. It is remarkable what mechanical skill is brought to bear upon these operations. In the first place, the plough cuts a furrow as straight and even as if it were turned by machinery. A kind of *esprit de corps* animates the ploughmen to a vigorous ambition in the work. They are trained to it with as much singleness of purpose as the smiths of Sheffield are to the forging of penknife blades. On a large estate like that occupied by Mr. Jonas, they constitute an order, not of Odd Fellows, but of Straight Furrowmen, and are jealous of the distinction. When the ground is well prepared, and made as soft, smooth, and even as a garden, the drilling process is performed with a judgement of the eye and skill of hand more marvellous still. The straightness of the lines of verdure which, in a few weeks, mark the tracks of the seed-tubes, is surprising. They are drawn and graded with such precision that, when the plants are at a certain height, a horse-hoe, with eight blades, each wide enough to cut the whole intervening space between two rows, is passed, hoeing four or five drills at once. Of course, if the lines of the drill and hoe did not exactly correspond with each other, whole rows of turnips would be cut up and destroyed. I saw this process going on in a turnip field, and thought it the most skilful operation connected with agriculture that I had ever witnessed.

One of the principal advantages Mr. Jonas realises in cultivating such an extent of territory, is the ability to economise his working forces, of man, beast, and agricultural machinery. He saves what may be called the superfluous fractions, which small farmers frequently lose. For instance, a man with only fifty acres would need a pair of stout horses, a plough, cart, and all the other implements necessary for the growth and gathering of the usual crops. Now, Mr. Jonas has proved by experience, that, in cultivating his great occupation, the average force of two and a quarter horses is sufficient for a hundred acres. Here is a saving of almost one half the expense of horse-force per acre which the small farmer incurs, and full one half of the use of carts, ploughs, and other implements. The whole number of horses employed is about seventy-six; and the number of men and boys about a hundred. The whole of this great force is directed by Mr. Jonas and his sons with as much apparent ease and equanimity as the captain of a Cunarder would manifest in guiding a steamship across the Atlantic. The helm and ropes of the establishment obey the motion of one mind with the same readiness and harmony.

A fact or two may serve an American farmer as a tangible measure whereby to estimate the extent of the operations thus conducted by one man. To come up to the standard of scientific and successful agriculture in England, it is deemed requisite that a tenant farmer, on renting an occupation, should have capital sufficient to invest £10, or $50, per acre in stocking it with cattle, sheep, horses, farming implements, fertilisers, etc. Mr. Jonas, beyond a doubt, invests capital after this ratio upon the estate he tills. If so, then the total amount appropriated to the land which he *rents* cannot be less than £30,000, or nearly $150,000. The inventory of his live stock, taken at last Michaelmas, resulted in these figures:—Sheep, £6,581; horses, £2,487; bullocks, £2,218; pigs, £452; making a grand total of £11,638. Every animal bred on the estate is fatted, but by no means with the grain and roots grown upon it. The outlay for oil-cake and corn purchased for feeding, amounts to about £4,000 per annum. Another heavy expenditure is about £1,700 yearly for artificial fertilisers, consisting of guano and blood-manure. Mr. Jonas is one of the directors of the company formed for the manufacture of the latter.

The whole income of this establishment is realised from two sources—meat and grain. And this is the distinguishing characteristic of English farming generally. Not a pound of hay, straw, or roots is sold off the estate. Indeed, this is usually prohibited by the conditions of the contract with the landlord. So completely has Mr. Jonas adhered to this rule, that he could not give me the market price of hay, straw, or turnips per ton, as he had never sold any, and was not in the habit of noticing the market quotations of those products. I was surprised at one fact which I learned in connection with his economy. He keeps about 170 bullocks; buying in October and selling in May. Now, it would occasion an American farmer some wonderment to be told that this great herd of cattle is fed and fatted almost entirely for the manure they make. It is doubtful if the difference between the cost and selling prices averages £2, or $10, per head. For instance, the bullocks bought in will average £13 or £14. A ton of linseed-cake and some meal are given to each beast before it is sent to market, costing from £10 to £12.

When sold, the bullocks average £24 or £25. Thus the cake and the meal equal the whole difference between the buying and selling price, so that all the roots, chaff, and attendance go entirely to the account of manure. These three items, together with the value of pasturage for the months the cattle may lie in the fields, from October to May inclusive, could hardly amount to less than £5 per beast, which, for 170, would be £850. Then £1,700 are paid annually for guano and artificial manures. Now add the value of the wheat, oat and barley straw grown on 1,500 acres, and mostly thrown into the barn-yards, or used as bedding for the stables, and you have one great division of the fertilising department of Chrishall Grange. The amount of these three items cannot be less than £3,000. Then there is another source of fertilisation nearly as productive and valuable. Upwards of 3,000 sheep are kept on the estate, of which 1,200 are breeding ewes. These are folded, acre by acre, on turnips, cole, or trefoil, and those fattened for the market are fed with oil-cake in the field. The locusts of Egypt could not have left the earth barer of verdure than these sheep do the successive patches of roots in which they are penned for twenty-four or forty-eight hours, nor could any other process fertilise the land more thoroughly and cheaply. Then 76 horses and 200 fattening hogs add their contingent to the manurial expenditure and production of the establishment. Thus the fertilising material applied to the estate cannot amount to less than £5,000, or $24,000, per annum.

Sheep are the most facile and fertile source of nett income on the estate. Indeed, nearly all the profit on the production of meat is realised from them. Most of those I saw were Southdowns and Hampshires, pure or crossed, with here and there a Leicester. After being well fattened, they fetch in the market about double the price paid for them as stock sheep. About 2,000, thus fattened, including lambs, are sold yearly. They probably average about £2, or $10, per head; thus amounting to the nice little sum of £4,000 a year, as one of the

sources of income.

Perhaps it would be easier to estimate the total expenditure than the gross income of such an establishment as that of Mr. Jonas. We have aggregated the former in a lump; assuming that the whole capital invested in rent, live stock, agricultural machinery, manures, labor of man and horse, fattening material, etc. , amounts to £30,000. We may extract from this aggregate several estimated items which will indicate the extent of his operations, putting the largest expenditure at the head of the list.

Corn and oil-cake purchased for feeding £4,000
Guano and manufactured manures 1,700
Labor of 100 men and boys at the average of £20 per annum 2,000
Labor of 76 horses, including their keep, £20 per annum 1,500
Use and wear of steam-engine and agricultural machinery 500
Commutation money to men for beer 400

--

£10,100

These are some of the positive annual outlays, without including rent, interest on capital invested, and other items that belong to the debit side of the ledger. The smallest on the list given I would commend to the consideration of every New England farmer who may read these pages. It is stated under the real fact. The capacity of English laborers for drinking strong beer is a wonder to the civilised world. They seem to cling to this habit as to a vital condition of their very life and being. One would be tempted to think that malt liquor was a primary and bread a secondary necessity to them; it must cost them most of the two, at any rate. And generally they are as particular about the quality as the quantity, and complain if it is not of "good body," as well as full tale. In many cases the farmer furnishes it to them; sometimes brewing it himself, but more frequently buying it already made. Occasionally a farmer "commutes" with his men; allowing a certain sum of money weekly in lieu of beer, leaving them to buy and use it as they please. I understood that Mr. Jonas adopts the latter course, not only to save himself the trouble of furnishing and rationing such a large quantity of beer, but also to induce the habit among his men of appropriating the money he gives them instead of drink to better purposes. The sum paid to them last year was actually £452, or about $2,200! Now, it would be quite safe to say, that there is not a farm in the State of Connecticut that produces pasturage, hay, grain, and roots enough to pay this beer-bill of a single English occupation! This fact may not only serve to show the scale of magnitude which agricultural enterprise has assumed in the hands of such men as Mr. Jonas, but also to indicate to our American farmers some of the charges upon English agriculture from which they are exempt; thanks to the Maine Law, or, to a better one still, that of voluntary disuse of strong drink on our farms. I do not believe that 100 laboring men and boys could be found on one establishment in Great Britain more temperate, intelligent, industrious, and moral than the set employed by Mr. Jonas. Still, notice the tax levied upon his land by this beer-impost. It amounted last year to three English shillings, or seventy-two cents, on every acre of the five consolidated farms, including all the space occupied by hedges, copses, buildings, etc. Suppose a Maine farmer were obliged, by an inexorable law of custom, to pay a beer-tax of seventy-two cents per acre on his estate of 150 acres, or $108, annually, would he not be glad to "commute" with his hired men, by leaving them in possession of his holding and migrating to some distant section of the country where such a custom did not exist?

The gross income of this great holding it would be more difficult to estimate. But no one can doubt the yearly issues of Mr. Jonas' balance-sheet, when he has been able to expand his operations gradually to their present magnitude from the capital and experience acquired by successful farming. Perhaps

the principal sources of revenue would approximate to the following figures:—

2,000 fat sheep and lambs at £2 £4,000
   150 fat bullocks at £25   3,750
   200 fat pigs = 40,000 lbs., at 4d 666
22,500 bushels of wheat, at 6s 6,750
9,375 bushels of oats, at 2s 937
7,500 bushels of barley, at 3s 1,125

Total of these estimated items £17,228

This, of course, is a mere estimate of the principal sources of income upon which Mr. Jonas depends for a satisfactory result of his balance-sheet. Each item is probably within the mark. I have put down the crop of wheat of 750 acres at the average of thirty bushels per acre, and at 6s. per bushel, which are quite moderate figures. I have assumed 375 acres each for barley and oats, estimating the former at forty bushels per acre, and the latter at fifty; then reserving half of the two crops for feeding and fatting the live stock; also all the beans, peas, and roots for the same purpose. If the estimate is too high on some items, the products sold, and not enumerated in the foregoing list, such as cole and other seeds, will rectify, perhaps, the differences, and make the general result presented closely approximate to the real fact.

As there is probably no other farm in Great Britain of the same size so well calculated to test the best agricultural science and economy of the day as the great occupation of Mr. Jonas, and as I am anxious to convey to American farmers a well-developed idea of what that science and economy are achieving in this country, I will dwell upon a few other facts connected with this establishment. The whole space of 3,000 acres is literally under cultivation, or in a sense which we in New England do not generally give to that term—that is, there is not, I believe, a single acre of permanent meadow in the whole terri-

tory. All the vast amount of hay consumed, and all the pasture grasses have virtually to be grown like grain. There is so much ploughing and sowing involved in the production of these grass crops, that they are called "seeds." Thus, by this four-course system, every field passes almost annually under a different cropping, and is mowed two or three times in ten years. This fact, in itself, will not only suggest the immense amount of labor applied, but also the quality and condition of 3,000 acres of land that can be surfaced to the scythe in this manner.

The *seeds* or grasses sown by Mr. Jonas for pasturage and hay are chiefly white and red clover and trefoil. His rule of seeding is the following:—

Wheat, from   8 to 10 pecks per acre
Barley, from   12 to 14   "   "   "
Oats, from   18 to 22   "   "   "
Winter Beans,   8   "   "   "
Red Clover,   20 lbs   "   "
White Clover, 16 lbs   "   "
Trefoil,   30 to 35 lbs. "   "

This, in New England, would be called very heavy seeding, especially in regard to oats and the grasses. I believe that twelve pecks of oats to the acre, rather exceed our average rule. Good clover seed should weigh two pounds to the quart, and eight quarts, or sixteen pounds, are the usual seeding with us.

As labor of horse and man must be economised to the best advantage on such an estate, it may be interesting to know the expense of the principal operations. The cost of ploughing averages 7s. 6d., or $1 80c. per acre. For roots, the land is ploughed three or four times, besides harrowing, drilling, and rolling. The hoeing of wheat and roots varies from 2s. to 5s., or from 48c. to $1 20c. per acre.

The sheep are all folded on turnips or grass fields, except the breeding ewes in the lambing season. The enclosures are made of *hurdles*, of which all reading Americans have read, but not one in a thousand ever has seen. They are a kind of diminutive, portable, post-and-rail fence, of the New England pattern, made up in permanent *lengths*, so light that a stout man might carry two or three of them on his shoulders at once. The two posts are sawed or split pieces of wood, about two inches thick, three wide, and from five to six feet in length. They are generally square-morticed for the rails, which are frequently what we should call split hoop-holes, but in the best kind are slats of hard wood, about two and a half inches wide and one in thickness. Midway between the two posts, the rails are nailed to an upright slat or brace, to keep them from swaying. Sometimes a farmer makes his own hurdles, thus furnishing indoor work for his men in winter, when they cannot labor in the fields; but most generally they are bought of those who manufacture them on a large scale. Some idea of the extent of sheep-folding on Chrishall Grange may be inferred from the fact, that the hurdling on it, if placed in one straight, continuous line, would reach full ten miles!

A portable steam engine, of twelve-horse power, looking like a common railway locomotive strayed from its track and taken up and housed in a farmer's waggon-shed, performs prodigies of activity and labor. Indeed, search the three realms through and through, and you would hardly find one on its own legs doing such remarkable varieties of work. Briareus, with all his fabled faculties, never had such numerous and supple fingers as this creature of human invention. When set a-going, they are clattering and whisking and frisking everywhere, on the barn-floor, on the hay-loft, in the granary, under the eaves, down cellar, and all this at the same time. It is doubtful if any stationary engine in a machine shop ever performed more diversified operations at once; thus proving most conclusively how a farmer may work motive power which it was once thought preposterous in him to think of using. It threshes wheat and other kinds of grain at the rate of from 400 to 500 bushels a day; it conveys the straw up to a platform across what we call the "*great beams*," where it is cut into chaff and dropped into a great bay, at the trifling expense of sixpence, or twelve cents, per quantity grown on an

acre! While it is doing this in one direction, it is turning machinery in another that cleans and weighs the grain off into sacks ready for the market. Open the doors right and left and you find it at work like reason, breaking oil-cake, grinding corn for the fat stock, turning the grindstone, pitching, pounding, paring, rubbing, grabbing, and twisting, threshing, wrestling, chopping, flopping, and hopping, after the manner of "The Waters of Lodore."

The housings for live stock are most admirably constructed as well as extensive, and all the great yards are well fitted for making and delivering manure. I noticed here the best arrangement for feeding swine that I had ever seen before, and of a very simple character. Instead of revolving troughs, or those that are to be pulled out like drawers to be cleaned, a long, stationary one, generally of iron, extends across the whole breadth of the compartment next to the feeding passage. The board or picket-fence forming this end of the enclosure, from eight to twelve feet in length, is hung on a pivot at each side, playing in an iron ring or socket let into each of the upright posts that support it. Midway in the lower rail of this fence is a drop bolt which falls into the floor just behind the trough. At the feeding time, the man has only to raise this bolt and let it fall on the inner side, and he has the whole length and width of the trough free to clear with a broom and to fill with the feed. Then, raising the bolt, and bringing it back to its first place, the operation is performed in a minute with the greatest economy and convenience.

There was one feature of this great farm home which I regarded with much satisfaction. It was the housing of the laborers employed on the estate. This is done in blocks of well-built, well-ventilated, and very comfortable cottages, all within a stone's throw of the noble old mansion occupied by Mr. Jonas. Thus, no long and weary miles after the fatigue of the day, or before its labor begins, have to be walked over by his men in the cold and dark, as in many cases in which the agricultural laborer is obliged to trudge on foot from a distant village

to his work, making a hard and sunless journey at both ends of the day.

Although my visit at this, perhaps the largest, farming establishment in England, occupied only a few hours, I felt on leaving that I had never spent an equal space of time more profitably and pleasantly in the pursuit or appreciation of agricultural knowledge. The open and large-hearted hospitality and genial manners of the proprietor and his family seemed to correspond with the dimensions and qualities of his holding, and to complete, vitalise, and beautify the symmetries of a true ENGLISH FARMER'S HOME.

CHAPTER X.

ROYSTON AND ITS SPECIALITIES— ENTERTAINMENT IN A SMALL VILLAGE—ST. IVES—VISITS TO ADJOINING VILLAGES—A FEN-FARM— CAPITAL INVESTED IN ENGLISH AND AMERICAN AGRICULTURE COMPARED—ALLOTMENTS AND GARDEN TENANTRY—BARLEY GROWN ON OATS.

From Chrishall Grange I went on to Royston, where I found very quiet and comfortable quarters in a small inn called "The Catherine Wheel," for what reason it is not yet clear to my mind, and the landlady could not enlighten me on the subject. I have noticed two inns in London of the same name, and have seen it mounted on several other public houses in England. Why that ancient saint and the machinery of her torture should be alone selected from the history and host of Christian martyrs, and thus associated with houses of entertainment for man and beast, is a mystery which I will not undertake to explore. To be sure, the head of a puncheon of rum is round like a wheel, and if the liquor were not too much diluted with water, it might make a revolving illumination quite interesting, if set on fire and rolled into the gutter. It may possibly suggest that lambent ignition of the brain which the fiery drinks of the establishment produce, and which so many infatuated victims think delightful. Both these inferences, and all others I could fancy, are so dubious that I will not venture further into the meaning of this singular appellation given to a tavern.

Royston is a goodly and comfortable town, just inside the eastern boundary of Hertfordshire. It has its full share of half-legible and interesting antiquities, including the ruins of a royal palace, a cave, and several other broken monuments of the olden time, all festooned with the web-work of hereditary fancies, legends, and shreds of unravelled history dyed to the vivid colors of variegated imagination. It also boasts and enjoys a great, breezy common, large enough to hold such another town, and which few in the kingdom can show. Then, if it cannot cope with Glastonbury in showing, to the envious and credulous world, a thorn-tree planted by Joseph of Arimathæa, and blossoming always at Christmas, it can fly a bird of greater antiquity, which never flapped its wings elsewhere, so far as I can learn. It may be the lineal descendant of Noah's raven that has come down to this particular community without a cross with any other branch of the family. It is called "The Royston Crow," and is a variety of the genus which you will find in no other country. It is a great, heavy bird, larger than his colored American cousin, and is distinguished by a white back. Indeed, seen walking at a distance, he looks like our Bobolink expanded to the size of a large hen-hawk. To have such a wild bird all to themselves, and of its own free will, notwithstanding the length and power of its wings, and the force of centrifugal attractions, is a distinction which the good people of this favored town have good reason to appreciate at its proper value. Nor are they insensible to the honor. The town printer put into my hands a monthly publication called "THE ROYSTON CROW," containing much interesting and valuable information. It might properly have embraced a chapter on entomology; but, perhaps, it would have been impolitic for the personal interests of the bird to have given wide publicity to facts in this department of knowledge. For, after all, there may exist in the neighborhood certain special kinds of bugs and other insects which lie at the foundation of his preference for the locality.

The next day I again faced northward, and walked as far as Caxton, a small, rambling village, which looked as if it had not shaved and washed its face, and put on a clean shirt for a shocking length of time. It was dark when I reached it; having walked twelve miles after three p.m. There was only one inn, properly speaking, in the town, and since the old coaching time, it had contracted itself into the fag-end of a large, dark, seedy-looking building, where it lived by selling beer and other sharp and cheap drinks to the villagers; nineteen-twentieths of whom appeared to be agricultural laborers. The entertainment proffered on the sign-board over the door was evidently limited to the tap-room. Indeed, this and the great, low-jointed and brick-floored kitchen opening into to it, seemed to constitute all the living or inhabited space in the building. I saw, at a glance, that the chance for a bed was faint and small; and I asked Landlord Rufus for one doubtingly, as one would ask for a ready-made pulpit or piano at a common cabinet-maker's shop. He answered me clearly enough before he spoke, and he spoke as if answering a strange and half-impertinent question, looking at me searchingly, as if he suspected I was quizzing him. His "No!" was short and decided; but, seeing I was honest and earnest in the inquiry, he softened his negative with the explanation that their beds were all full. It seemed strange to me that this should be so in a building large enough for twenty, and I hesitated hopefully, thinking he might remember some small room in which he might put me for the night. To awaken a generous thought in him in this direction, I intimated how contented I would be with the most moderate accommodation. But it was in vain. The house was full, and I must seek for lodging elsewhere. There were two or three other public houses in the village that might take me in. I went to them one by one. They all kept plenty of beer, but no bed. They, too, looked at me with surprise for asking for such a thing. Apparently, there had been no demand for such entertainment by any traveller since the stage-coach ceased to run through the village. I went up and down, trying to negotiate with the occupants of some of the best-looking cottages for a cot or bunk; but they had none to spare, as the number of wondering children that stared at me kindly, at once suggested before I put the question.

It was now quite dark, and I was hungry and tired; and the prospect of an additional six miles walk was not very animating. What next? I will go back to Landlord Rufus and try a new influence on his sensibilities. Who knows but it will succeed? I will touch him on his true character as a Briton. So I went back, with my last chance hanging on the experiment. I told him I was an American traveller, weary, hungry, and infirm of health, and would pay an extra price for an extra effort to give me a bed for the night. I did not say all this in a *Romanus-civus-sum* sort of tone. No! dear, honest Old Abe, you would have done the same in my place. I made the great American Eagle coo like a dove in the request; and it touched the best instincts of the British Lion within the man. It was evident in a moment that I had put my case in a new aspect to him. He would talk with the "*missus;*" he withdrew into the back kitchen, a short conference ensued, and both came out together and informed me that they had found a bed, unexpectedly vacant, for my accommodation. And they would get up some tea and bread and butter for me, too. Capital! A sentiment of national pride stole in between every two feelings of common satisfaction at this result. The thought would come in and whisper, not for your importunity as a common fellow mortal were this bed and this loaf unlocked to you, but because you were an American citizen.

So I followed "the missus" into that great kitchen, and sat down in one corner of the huge fire-place while she made the tea. It was a capacious museum of culinary curiosities of the olden time, all arranged in picturesque groups, yet without any aim at effect. Pots, kettles, pans, spits, covers, hooks and trammels of the Elizabethan period, apparently the heirlooms of several intersecting generations, showed in the fire-light like a work of artistry; the sharp, silvery brightness of the tin and the florid flush of burnished copper making distinct disks in the darkness. It was with a rare sentiment of comfort that I sat by that fire of crackling faggots, looked up at the stars that dropped in their light as they passed over the top of the great chimney, and glanced around at the sides of that old English kitchen, panelled with plates and platters and dishes of all sizes and uses. And this fire was kindled and this tea-kettle was singing for me really because I was an American! I could not forget that—so I deemed it my duty to keep up the character. Therefore, I told the *missus* and her bright-eyed niece a great many stories about America; some of which excited their admiration and wonder. Thus I sat at the little, round, three-legged table, inside the out-spreading chimney, for an hour or more, and made as cozy and pleasant a meal of it as ever I ate. Besides all this, I had the best bed in the house, and several "Good nights!" on retiring to it, uttered with hearty good-will by voices softened to an accent of kindness. Next morning I was introduced into the best parlor, and had a capital breakfast, and then resumed my walk with a pleasant memory of my entertainment in that village inn.

I passed through a fertile and interesting section to St. Ives, in Huntingdonshire. Here I remained with some friends for a week, visiting neighboring villages by day and returning at night. St. Ives is a pleasant, well-favored town, just large enough to constitute a coherent, neighborly, and well-regulated community. It is the centre-piece of a rich, rural picture, which, without any strikingly salient features, pleases the eye with lineaments of quiet beauty symmetrically developed by the artistry

of Nature. The river Ouse meanders through a wide, fertile flat, or what the Scotch would call a strath, which gently rises on each side into pleasantly undulating uplands. Parks, groves, copses, and hedge-row trees are interspersed very happily, and meadow, pasture, and grain-fields seen through them, with villages, hamlets, farm-houses, and isolated cottages, make up a landscape that grows more and more interesting as you contemplate it. And this placid locality, with its peaceful river seemingly sleeping in the bosom of its long and level meadows, was the scene of Oliver Cromwell's young, fiery manhood. Here, where Nature invites to tranquil occupations and even exercises of the mind, he trained the latent energies of his will for action in the great drama that overturned a throne and transformed a nation. Here, till very lately, stood his "barn," and here he drilled the first squadron of his "Ironsides."

My friend and host drove me one day to see a fen-farm a few miles beyond Ramsey, at which we remained over night and enjoyed the old-fashioned English hospitality of the establishment with lively relish. It was called "The Four-Hundred-Acre-Farm," to distinguish it from a hundred others, laid out on the same dead level, with lines and angles as straight and sharp as those of a brick. You will meet scores of persons in England who speak admiringly of the great prairies of our Western States—but I never saw one in Illinois as extensive as the vast level expanse you may see in Cambridgeshire and Lincolnshire. In fact, the space of a large county has been fished up out of a shallow sea of salt water by human labor and capital. I will not dwell here upon the expense, process, and result of this gigantic operation. It would require a whole chapter to convey an approximate idea of the character and dimensions of the enterprise. The feat of Cyrus in turning the current of the Euphrates was the mere making of a short mill-race compared with the labor of lifting up these millions of acres bodily out of the flood that had covered and held them in quiescent solution since

the world began.

This Great Prairie of England, generally called here the Fens, or Fenland, would be an interesting and instructive section for the agriculturists of our Western States to visit. They would see how such a region can be made quite picturesque, as well as luxuriantly productive. Let them look off upon the green sea from one of the upland waves, and it will be instructive to them to see and know, that all the hedge-trees, groves, and copses that intersect and internect the vast expanse of green and gold were planted by man's hands. Such a landscape would convince them that the prairies of Illinois and Iowa may be recovered from their almost depressing monotony by the same means. The soil of this district is apparently the same as that around Chicago—black and deep, on a layer of clay. It pulverises as easily in dry weather, and makes the same inky and sticky composition in wet. To give it more body, or to cross it with a necessary and supplementary element, a whole field is often trenched by the spade as clean as one could be furrowed by the plough. By this process the substratum of clay is thrown up, to a considerable thickness, upon the light, black, almost volatile soil, and mixed with it when dry; thus giving it a new character and capacity of production.

Everything seems to grow on a Californian scale in this fen district. Although the soil thus rescued from the waters that had flooded and half dissolved it, was at first as deep, black, and naturally fertile as that of our prairies, those who commenced its cultivation did not make the same mistake as did our Western farmers. They did not throw their manure into the broad draining canals to get rid of it, trusting to the inexhaustible fertility of the alluvial earth, as did the wheat growers of Indiana and Illinois to their cost; but they husbanded and well applied all the resources of their barn-yards. In consequence of this economy, there is no deterioration of annual averages of their crops to be recorded, as in some of our prairie States, which have been boasting of the natural and inexhaustible fertility

ERCHANT

of their soil even with the record of retrograde statistics before their eyes. The grain and root crops are very heavy; and a large business is done in growing turnip seed for the world in some sections of this fen country. A large proportion of the quantity we import comes from these low lands.

Our host of the Four-Hundred-Acre Farm took us over his productive occupation, which was in a very high state of cultivation. The wheat was yellowing to harvest, and promised a yield of forty-two bushels to the acre. The oats were very heavy, and the root crops looked well, especially a field of mangel-wurzel. He apportions his land to different crops after this ratio:— Wheat, 120 acres; oats, 80; rye-grass and clover, 50; roots, 60. His live stock consisted of 300 sheep, 50 to 60 head of cattle, and 70 to 80 hogs. His working force was from 10 to 12 men, 14 farm horses, and 4 nags. It may interest some of my American readers to know the number, character, and cost of the implements employed by this substantial English farmer in cultivating an estate of 400 acres. I noted down the following list, when he was showing us his tool-house:—

	£	$	£	$
6 Ploughs	at 4 each = 20	24		
= 120				
6 Horse-carts,	at 14 each = 70			
84 = 420				
1 Large Iron Roller and Gearing,				
13 = 65				
1 Cambridge Roller		14		
= 70				
1 Twelve-Coulter Drill		46		
= 230				
3 Harrows	at 3 each = 15	9		
= 45				
2 Great Harrows	at 3 each =			
15 6 = 30				

--- ---

Total cost of these Implements £196 $980

These figures will represent the working forces and implemental machinery of a well-tilled farm of 400 acres in England. They will also indicate the amount of capital required to cultivate an estate of this extent here. Let us compare it with the amount generally invested in New England for a farm of equal size. Thousands that have been under cultivation for a hundred years, may be bought for £5, or $25, per acre, including house, barn, and other buildings and appurtenances. It is a very rare thing for a man with us to buy 400 acres at once; but if he did, it would probably be on these conditions:— He would pay £400, or $2,000, down at the time of purchase, giving his notes for the remaining £1,600, or $8,000, at 6 per cent. interest payable annually, together with the yearly instalment of principal specified in each note. He would perhaps have £200, or $1,000, left of his capital for working power and agricultural implements. He would probably divide it after the following manner:—

	£	£	$
2 Yokes of Oxen, at			25
= 50 = 250			
1 Horse		20 =	
100			
2 Ox-carts, at		15 =	
30 = 150			
1 Waggon		20	
= 100			
2 Ox-sleds, at			1 = 2
= 10			
2 Ox-ploughs, at			2 =
4 = 20			
1 Single Horse-plough		1 =	5
2 Harrows			2 = 4
= 20			

Cradles, scythes, hoes, rakes, flails, etc.
4 = 20
Fanning-mill, hay-cutter, and corn-sheller. 4 = 20
15 Cows, steers, and heifers 45 = 225
6 Shoats, or pigs, six months old 10 = 50

These figures would indicate a large operation for a practical New England farmer, who should undertake to purchase and cultivate an estate of 400 acres. Indeed, not one in a hundred buying such a large tract of land would think of purchasing all the implements on this list at once, or entirely new. One of his carts, sleds, and harrows would very probably be "second-handed," and bought at half the price of a new one. Thus, a substantial farmer with us would think he was beginning on a very satisfactory and liberal footing, if he had £200, or $1,000, in ready money for stocking a holding of 400 acres with working cattle and implemental machinery, cows, pigs, etc. Now, compare this outlay with that of our host of the Four-Hundred-Acre Farm in Lincolnshire. We will begin with his—

	£	£	$
14 Farm horses, at the low figure of			
20 each = 280 = 1,400			
4 Nags, or saddle and carriage horses 20 each = 80 = 400			
300 Stock sheep 1 each = 300 = 1,500			
70 Pigs, of different ages 2 each = 140 = 900			
50 Head of cattle (cows, bullocks, etc.) 12 each = 600 = 3,000			
Carts, drills, rollers, ploughs and other implements 1,000 = 5,000			

£2,400 $12,200

The average rent of such land in England must be at least £1 10s. per acre, and the tenant farmer must pay half of this out of the capital he begins with; which, on 400 acres, would amount to £300. Then, if he buys a quantity of artificial manures equal to the value of 10s. per acre, he will need to expend in this department £200. Next, if he purchases corn and oil-cake at the same ratio for his cattle and sheep as that adopted by Mr. Jonas, of Chrishall Grange, he will want £1,000 for his 60 head of cattle and 300 sheep. In addition to these items of expenditure, he must pay his men weekly; and the wages of ten, at 10s. per week, for six months, amount to £130. Add an economical allowance for family expenses for the same length of time, and for incidental outgoes, and you make up the aggregate of £4,000, which is £10 to the acre, which an English farmer needs to have and invest on entering upon the cultivation of a farm, great or small. This amount, as has been stated elsewhere, is the rule for successful agriculture in this country.

These facts will measure the difference between the amounts of capital invested in equal spaces of land in England and America. It is as ten to one, assuming a moderate average. Here, a man would need £1,500, or more than $7,000, to begin with on renting a farm of 150 acres, in order to cultivate it successfully. In New England, a man would think he began under favorable auspices if he were able to enter upon the occupancy of equal extent with £100, or about $500.

On returning from the Fens, I passed the night and most of the following day at Woodhurst, a village a few miles north of St. Ives, on the upland rising gently from the valley of the Ouse. My host here was a farmer, owning the land he tilled, cultivating it and the moral character and happiness of the little community, in which he moved as a father, with an equally generous heart and hand, and reaping a liberal reward from both departments of his labor. He took me over his fields, and showed me his crops and live stock, which were in excellent condition. Harvesting had already commenced, and the reapers were at work, men and women, cutting wheat and barley. Few of them used sickles, but a curved knife, wider than the sickle, of nearly the same shape, minus the teeth. A man generally uses two of them. With the one in his left hand he gathers in a good sweep of grain, bends it downward, and with the other strikes it close to the ground, as we cut Indian corn. With the left-hand hook and arm, he carries on the grain from the inside to the outside of the swath or "work," making three or four strokes with the cutting knife; then, at the end, gathers it all up and lays it down in a heap for binding. This operation is called "bagging." It does not do the work so neatly as the sickle, and is apt to pull up many stalks by the roots with the earth attaching to them, especially at the last, outside stroke.

I was struck with the economy adopted by my host in loading, carting and stacking or ricking his grain. The operation was really performed like clockwork. Two or three men were stationed at the rick to unload the carts, two in the fields to load them, and several boys to lead them back and forth to the two parties. They were all one-horse carts, and so timed that a loaded one was always at the rick and an empty one always in the field; thus keeping the men at both ends fully employed from morning until night, pitching on and pitching off; while boys, at 6d. or 8d. a day, led the horses.

On passing through the stables and housings for stock, I noticed a simple, yet ingenious contrivance for watering cattle, which I am not sure I can describe accurately enough, without a drawing, to convey a tangible idea of it to my agricultural neighbors in America. It may be called the *buoy-cock*. In the first place, the water is brought into a cistern placed at one end of the stable or shed at a sufficient elevation to give it the necessary fall in all the directions in which it is to be conducted. The pipe used for each cow-box or manger connects each with the cistern, and the distributing end of it rests upon, or is suspended over, the trough assigned to each animal. About one-third of this trough, which was here a cast-iron box, about twelve inches deep and wide, protrudes through the boarding of the stable. In this outside compartment is placed a hollow copper ball attached to a lever, which turns the axle or pivot of the cock. Now, this little buoy, of course, rises and falls with the water in the trough. When the trough is full, the buoy rises and raises the lever so as to shut off the water entirely. At every sip the animal takes, the buoy descends and lets on again, to a drop, a quantity equal to that abstracted from the inside compartment. Thus the trough is always kept full of pure water, without losing a drop of it through a waste-pipe or overflow. Where a great herd of cattle and a drove of horses have to be supplied from a deep well, as in the case of Mr. Jonas, at Chrishall Grange, this *buoy-cock* must save a great amount of labor.

I saw also here in perfection that garden allotment system which is now coming widely into vogue in England, not only adjoining large towns like Birmingham, but around small villages in the rural districts. It is well worthy of being introduced in New England and other states, where it would work equally well in various lines of influence. A landowner divides up a field into allotments, each generally containing a rood, and lets them to the mechanics, tradespeople and agricultural laborers of the town or village, who have no gardens of their own for the growth of vegetables. Each of these is better than a savings-bank to the occupant. He not only deposits his odd pennies but his odd hours in it; keeping both away from the public-house or from places and habits of idleness and dissipation. The days of Spring and Summer here are very long, and a man can see to work in the field as early as three o'clock in the morning, and as late as nine at night. So every journeyman blacksmith, baker or shoemaker may easily find four or five hours in the twenty-four for work on his allotment, after having completed the task or time due to his employer. He generally keeps a pig, and is on the *qui vive* to make and collect all the manure he can for his little farm. A field of several acres, thus divided and cultivated in allotments, presents as striking a combination of colors as an Axminster carpet. As every rood is subdivided into a great variety of vegetables, and as forty or fifty of such patches, lying side by side, present, in one *coup d'œil*, all the alternations of which these crops and colors are susceptible, the effect is very picturesque.

My Woodhurst friend makes his allotment system a source of much social enjoyment to himself and the poor villagers. He lets forty-seven patches, each containing twenty poles. Every tenant pays 10s., or $2 40c., annual rent for his little holding, Mr. E. drawing the manure for each, which is always one good load a year. Here, too, these little spade-farmers are put under the same *regime* as the great tenant agriculturists of the country. Each must farm his allotment according to the terms of the yearly lease. He must dig up his land with spade or pick, not plough it; and he is not allowed to work on it upon the Sabbath. But encouragements greatly predominate over restrictions, and stimulate and reward a high cultivation. *Eight* prizes are offered to this end, of the following amounts:—10s., 7s. 6d., 5s., 4s., 3s., 2s. 6d., 2s. and 1s. Every one who competes must not have more than half his allotment in potatoes. The greater the variety of vegetables the other half contains, the better is his chance for the first prize. The appraiser is some disinterested person of good judgment, perhaps from an adjoining town, who knows none of the competitors. To prevent any possible favoritism, the allotments are all numbered, and he awards prizes to numbers only, not knowing to whom they belong. Another feature, illustrating the generous disposition of the proprietor, characterises this good work. On the evening appointed for paying the rents, he gets up a regular, old-fashioned English supper of roast beef and plum-pudding for them, giving each fourpence instead of beer, so that they may all go home sober as well as cheerful. To see him preside at that table, with his large, round, rosy face beaming upon them with the quiet benevolence of a good heart, and to hear the fatherly and neighborly talks he makes to them, would be a picture and preaching which might be commended to the farmers of all countries.

I saw also a curious phenomenon in the natural world on this farm, which perhaps will be regarded as a fiction of fancy by many a reader. It was a large field of barley grown from *oats*! We have recently dwelt upon some of the co-workings of Nature and Art in the development of flowers and of several useful plants. But here is something stranger still, that seems to diverge from the line of any law hitherto known in the vegetable world. Still, for aught one can know at this stage of its action, it may be the same general law of development which we have noticed, only carried forward to a more advanced point of progress. I would commend it to the

deep and serious study of naturalists, botanists, or to those philosophers who should preside over the department of investigation to which the subject legitimately belongs. I will only say what I saw with my own eyes and heard with my own ears. Here, I repeat, was a large field of heavy grain, ready for harvest. The head and berry were *barley*, and the stalk and leaves were *oat*! Here, certainly, is a mystery. The barley sown on this field was the first-born offspring of oats. And the whole process by which this wonderful transformation is wrought, is simply this, and nothing more:—The oats are sown about the last week in June; and, before coming into ear, they are cut down within one inch and a half of the ground. This operation is repeated a second time. They are then allowed to stand through the winter, and the following season the produce is *barley*. This is the plain statement of the case in the very words of the originator of this process, and of this strange transmutation. The only practical result of it which he claims is this: that the straw of the barley thus produced is stouter, and stands more erect, and, therefore, less liable to be beaten down by heavy wind or rain. Then, perhaps, it may be added, this oat straw headed with barley is more valuable as fodder for live stock than the natural barley straw. But the value of this result is nothing compared with the issue of the experiment as proving the existence of a principle or law hitherto undiscovered, which may be applied to all kinds of plants for the use of man and beast. If any English reader of these notes is disposed to inquire more fully into this subject, I am sure he may apply without hesitation to Mr. John Ekins, of Bruntisham, near St. Ives, who will supply any additional information needed. He presented me with a little sample bag of this oat-born barley, which I hope to show my agricultural neighbors on returning to America.

CHAPTER XI.

THE MILLER OF HOUGHTON—AN HOUR IN HUNTINGDON—OLD HOUSES—WHITEWASHED TAPESTRY AND WORKS OF ART—"THE OLD MERMAID" AND "THE GREEN MAN"—TALK WITH AGRICULTURAL LABORERS—THOUGHTS ON THEIR CONDITION, PROSPECTS, AND POSSIBILITIES.

After a little more than a week's visit in St. Ives and neighboring villages, I again resumed my staff and set out in a westerly direction, in order to avoid the flat country which lay immediately northward for a hundred miles and more. Followed the north bank of the Ouse to Huntingdon. On the way, I stopped and dined with a gentleman in Houghton whose hospitality and good works are well known to many Americans. The locality mentioned is so identified with his name, that they will understand whom I mean. There was a good and tender-hearted man who lived in our Boston, called Deacon Grant; and I hope he is living still. He was so kind to everybody in trouble, and everybody in trouble went to him so spontaneously for sympathy and relief, that no one ever thought of him as belonging to a single religious congregation, but regarded him as Deacon of the whole of Boston—a kind of universal father, whose only children were the orphans and the poor men's sons and daughters of the city. The Miller of Houghton, as some of my readers will know, is just such another man, with one slight difference, which is to his advantage, as a gift of grace. He has all of Deacon Grant's self-diffusing life of love for his kind, generous and tender dispositions towards the poor and needy, and more than the Deacon's means of doing good; and, with all this, the indomitable energy and will and even the look of Cromwell. During my stay in the neighborhood, I was present at two large gatherings at his House of Canvas, with which he supplements his family mansion when the latter lacks the capacity of his heart in the way of accommodation. This tent, which he erects on his lawn, will hold a large congregation;

and, on both the occasions to which I refer, was well filled with men, women, and children from afar and near. The first was a re-union of the Sunday-school teachers and pupils of the county, to whom he gave a sumptuous dinner; after which followed addresses and some business transactions of the association. The second was the examination of the British School of the village, founded and supported, I believe, by himself. At the conclusion of the exercises, which were exceedingly interesting, the whole company, young and old, adjourned to the lawn, where the visitors and elder people of the place were served with tea and coffee under the tent.

Then came "The Children's Hour." They were called in from their games and romping on the lawn, and formed into a circle fifty feet in diameter. And here and now commenced an entertainment which would make a more interesting picture than the old Apsley House Dinner. The good deacon of the county, with several assistants, entered this charmed circle of boys and girls, all with eyes dilated and eager with expectation, and overlooked by a circular wall of elder people radiant with the spirit of the moment. The host, in his white hat and grey beard, led the way with a basket on his arm, filled with little cakes, called with us gingernuts. He was followed by a file of other men with baskets of nuts, apples, etc. It was a most hilarious scene, exhilarating to all the senses to look upon, either for young or old. He walked around the ring with a grand, Cromwellian step, sowing a pattering rain of the little cakes on the clean-shaven lawn, as a farmer would sow wheat in his field, broadcast, in liberal handfuls. Then followed in their order the nut-sowers, apple-sowers, and the sowers of other goodies. When the baskets were emptied, the circular space enclosed was covered with as tempting a spread of dainties as ever fascinated the eyes of a crowd of little people. For a whole minute, longer than a full hour of ordinary schoolboy enjoyments, they had to stand facing that sight, involuntarily attitudinising for the plunge. At

the end of that long minute, the signal sounded, and, in an instant, there was a scene in the ring that would have made the soberest octogenarian shake his sides with the laughter of his youth. The encircling multitude of youngsters darted upon the thickly-scattered delicacies like a flock of birds upon a field of grain, with patter, twitter and flutter, and a tremor and treble of little short laughs; small, eager hands trying in vain to shut fast upon a large apple and several ginger-nuts at one grasp; slippings and trippings, tousling of tresses and crushing of dresses; boys and girls higgledy-piggledy; caps and bonnets piggledy-higgledy; little, red-faced Alexanders looking half sad, because they had filled their small pocket-worlds and both hands with apples and nuts, and had no room nor holding for more; little girls, with broken bonnet-strings, and long, sunny hair dancing over their eyes, stretching their short fingers to grasp another goodie,—all this, with the merry excitement of fathers and mothers, elder brothers and sisters, and other spectators, made it a scene of youthful life and delight which would test the genius of the best painters of the age to delineate. And Sir Roger Coverley Cromwell, the author of all this entertainment, would make a capital figure in the group, taken just as he looked at that moment, with his face illuminated with the upshooting joy of his heart, like the clear, frosty sky of winter with the glow and the flush of the Northern Lights.

The good Miller of Houghton, having added stone to stone until his mills can grind all the wheat the largest county can grow, has recently handed over to his sons the great business he had built up to such magnitude, and retired, if possible, to a more active life of benevolence. One of his late benefactions was a gift of £3,000, or nearly $15,000, toward the erection of an Independent Chapel in St. Ives.

At Huntingdon, I took tea and spent a pleasant hour with the principal of a select school, kept in a large, dignified and comfortable mansion, once occupied by the poet Cowper. In the yard be-

hind the house there is a wide-spreading and prolific pear-tree planted by his hands. This, too, was one of the thousands of old, stately dwellings you meet with here and there, which have no beginning nor end that you can get at. Cowper lived and wrote in this, for instance; but who lived in it a century before he was born? Who built it? Which of the Two Roses did he mount on his arms? Or did he live and build later, and dine his townsman, the great Oliver, or was he loyal to the last to Charles the First? These are questions that come up, on going over such a building, but no one can answer them, and you are left to the wisdom of limping legends on the subject. The present occupant has an antiquarian *penchant*; so, a short time after he took possession of the house, he began to make explorations in the walls and wainscotings, as men of the same mind have done at Nineveh and Pompeii. Having penetrated a thick surface of white lava, or a layer of lime, put on with a brush "in an earlier age than ours," he came upon a gorgeous wall of tapestry, with inwoven figures and histories of great men and women, quite as large as life, and all of very florid complexion and luxurious costumes. He has already exhumed a great many square yards of this picturesque fabric, wrought in by-gone ages, and is continuing the work with all the zest and success of a fortunate archæologist. Now it is altogether probable, that Cowper, as he sat in one of those rooms writing at his beautiful rhymes, had not the slightest idea that he was surrounded by such a crowd of kings, queens, and other great personages, barely concealed behind a thin cloud of white-wash.

It may possibly be true, that a few beautiful, fair-haired heretics in love or religion have been stone-masoned up alive in the walls of abbeys or convents. Sir Walter Scott leaned to that belief, and perhaps had credible history for it. But if the trowel has slain its thousands, the whitewash swab has slain its ten thousands of innocents. Think of the furlongs of richly-wrought tapestry, full of sacred and profane history, and the furlongs of curiously-carved panels, wainscoting, and cornice that floppy, sloppy, vandal brush of pigs' bristles and pail of diluted lime have eclipsed and obliterated for ever, and not a retributive drop of the villainous mixture has fallen into the perpetrator's eye to "make his foul intent seem horrible!" Think of Christian kings of glorious memory, even Defenders of the Faith, with their fair queens, princes of the blood, and knights, noble and brave, all, in one still St. Bartholomew night of that soft, thin, white flood, buried from the sight of the living as completely as the Roman sentinel at his post by the red gulf-stream of Vesuvius! Still, we must not be too hard on these seemingly barbarous transactions. "Not in anger, not in wrath," nor in foolish fancy, was that dripping brush always lifted upon these works of art. Many a person of cultivated taste saw a time when he could say, almost with Sancho Panza, "blessings on the man who invented whitewash! It covers a tapestry, a carving, or a sculpture all over like a blanket;" like that one spoken of in Macbeth. England is just beginning to learn what treasures of art in old mansions, churches and cathedrals were saved to the present age by a timely application of that cheap and healthy fluid. For there was a time when stern men of iron will arose, who had no fear of Gothic architecture, French tapestry, or Italian sculpture before their eyes; who treated things that had awed or dazzled the world as "baubles" of vanity, to be put away, as King Josiah put away from his realm the graven images of his predecessors. And these men thought they were doing good service to religion by pushing their bayonets at the most delicate works of the needle, pencil and chisel; ripping and slitting the most elaborately wrought tapestry,— stabbing off the fine leaf, and vine-work from carved cornices and wainscoting, and mutilating the marble lace-work of the sculptor in the old cathedrals. The only way to save these choice things was to make them suddenly take the white veil from the whitewasher's brush. Thousands of them were thus preserved, and they are now being brought forth to the light again, after having been shut away from the eye of man for several centuries.

The school-house is still standing in Huntingdon, in good condition and busy occupation, in which Oliver Cromwell stormed the English alphabet and carried the first parallel of monosyllables at the point of the pen. The very form or bench of oak from which he mounted the breach is still occupied by boys of the same size and age, with the same number of inches between their feet and the floor which separated it from his. Had the photographic art been discovered in his day, we might have had his face and form as he looked when seated as a rosy-faced, light-haired boy in the rank and file of the youngsters gathered within those walls. What an overwhelming revelation it would have been to his young, honest and merry mind, if some seer, like him who told Hazael his future, could have given him a sudden glimpse of what he was to be and do in his middle manhood!

After tea, I continued my walk westward to a small, quiet, comfortable village, about five miles from Huntingdon, where I became the guest of "The Old Mermaid," who extended her amphibious hospitalities to all strangers wishing bed and board for the night. Both I received readily and greatly enjoyed under her roof, especially the former. Never did I occupy a bed so fringed with the fanciful artistries of dreamland. It was close up under the thatched roof, and it was the most easy and natural thing in the world for the fancies of the midnight hour to turn that thatching into hair, and to cheat my willing mind with the delusion that I was sleeping with the long, soft tresses of Her Submarine Ladyship wound around my head. It was a delightful vagary of the imagination, which the morning light, looking in through the little checker-work window, gently dispelled.

The next day I bent my course in a north-westerly direction, and passed through a very fertile and beautiful section. The scenery was truly delightful;—not grand nor splendid, but replete with quiet pictures that please the eye and touch the heart with a sense of gladness. The soft mosaic work of the gently rounded hills, or figures wrought in wheat, barley, oats, beans, turnips, and meadow and pasture land, and grouped into landscapes in endless alternation of lights and shades, and all this happy little world now veiled by the low, summer clouds, now flooded by a sunburst between them—all these lovely and changing sceneries made my walk like one through a continuous gallery of paintings.

Harvesting had commenced in real earnest, and the wheat-fields were full of reapers, some wielding the sickle, others the scythe. When I saw men and women bending almost double to cut their sheaves close to the ground, I longed to walk through a barley-field with one of our American cradles, and show them how we do that sort of thing. As yet I have seen no reaping machines in operation, and I doubt if they will ever come into such extensive use here as with us, owing to the abundance of cheap labor in this country. I saw on this day's walk the heaviest crop of wheat that I have noticed since I left London. It must have averaged sixty bushels to the acre for the whole field.

Late in the afternoon it began to rain; and I was glad to find shelter and entertainment at a comfortable village inn, under the patronage of "The Green Man," perhaps a brother or near relative of Mermadam my hostess that entertained me the preceding night. It was a unique old building, or rather a concrete of a great variety of buildings devoted to a remarkable diversity of purposes, including brewing, farming, and other occupations. The large, low, dark kitchen was flanked by one of the old-fashioned fire-places, with space for a large family between the jambs, and the hollow of the chimney ample enough to show one of the smaller constellations at the top of it in a clear night. A seat on the brick or stone floor before one of these kitchen fire-places is to me the focus of the home comforts of the house, and I always make for it mechanically. As the darkness drew on, several agricultural laborers drifted in, one after the other, until the broad, deep pavement of the hearth was lined by a row of them, quite fresh from their work. They were quiet, sober-looking men, and they spoke with subdued voices, without animation or excitement, as if the fatigue of the day and the general battle of life had softened them to a serious, pensive mood and movement. As they sat drying their jackets around the fire, passing successive mugs of the landlord's ale from one to the other, they grew more and more conversational; and, as I put in a question here and there, they gave me an insight into the general condition, aspects and prospects of their class which I had not obtained before. They were quite free to answer any questions relating to their domestic economy, their earnings, spendings, food, drink, clothing, housing and fuel, also in reference to their educational and religious privileges and habits.

It was now the first week of harvest; and harvest in England, in any one locality, covers the space of a full month, in ordinary weather. Then, as the season varies remarkably, so that one county is frequently a week earlier in harvesting than that adjoining it on the north, the work for the sickle is often prolonged from the middle of July to the middle of September. This is the period of great expectation as well as toil for the agricultural laborers. Every man, woman, and boy of them is all put under the stimulus of extra earnings through these important weeks. Even the laborers hired by the year have a full month given them for harvesting forty or fifty extra shillings under this stimulus. Nearly all the grain in England is cut for a certain stipulated sum per acre; and thousands of all ages, with sickle or scythe in hand, see the sun rise and set while they are at work in the field. In the field they generally breakfast, lunch, and dine; and when it is considered there is daylight enough for labor between half-past three in the morning to half-past eight at night, one may easily see how many of the twenty-four hours they may bend to their toil. The price for cutting and binding wheat is from 10s. to 14s. , or from $2 40c. to $3 36c. per acre, and 8s., or $1 92c. per acre for oats

and barley. The men who cut, bind, and shock by the acre generally have to find their own beer, and will earn from 24s. to 28s., or from $5 76c. to $6 72c. per week. The regular laborers frequently let themselves to their employers during the harvest month at from 20s. to 24s. per week, which is just about double their usual wages. In addition to this pay, they are often allowed two quarts of ale and two quarts of small beer per day; not the small beer of New England, made only of hops, ginger, and molasses; but a far more stimulating drink, quite equal to our German *lager*. This gallon of beer will cost the farmer about 10d., or 20c. Where the piece-work laborer furnishes his own malt liquor, it must cost him on an average about an English shilling, or twenty-four cents, a day.

Two or three of the men who formed the circle around the fire at The Green Man, had come to purchase, or pay for, a keg of beer for their harvest allowance. It was to me a matter of half-painful interest to see what vital importance they attached to a supply of this stimulant—to see how much more they leaned upon its strength and comfort than upon food. It was not in my heart to argue the question with them, or to seek to dispel the hereditary and pleasant illusion, that beer alone, of all human drinks, could carry them through the long, hot hours of toil in harvest. Besides, I wished to get at their own free thoughts on the subject without putting my own in opposition to them, which might have slightly restricted their full expression. Every one of them held to the belief, as put beyond all doubt or question by the experience of the present and all past generations, that wheat, barley and oats could not be reaped and ricked without beer, and beer at the rate of a gallon a day per head. Each had his string of proofs to this conviction terminating in a pewter mug, just as some poor people praying to the Virgin have a string of beads ending in a crucifix, which they tell off with honest hearts and sober faces. Each could make it stand to reason that a man could not bear the heat and burden of harvest la-

bor without beer. Each had his illustration in the case of some poor fellow who had tried the experiment, out of principle or economy, and had failed under it. It was of no use to talk of temperance and all that. It was all very nice for well-to-do people, who never blistered their hands at a sickle or a scythe, to tell poor, laboring men, sweating at their hot and heavy work from sun to sun, that they must not drink anything but milk and water or cold tea and coffee, but put them in the wheat-field a few days, and let them try their wishy-washy drinks and see what would become of them. As I have said, I did not undertake to argue the men out of this belief, partly because I wished to learn from them all they thought and felt on the subject, and partly, I must confess, because I was reluctant to lay a hard hand upon a source of comfort which, to them, holds a large portion of their earthly enjoyments, especially when I could not replace it with a substitute which they would accept, and which would yield them an equal amount of satisfaction.

A personal habit becomes a "second nature" to the individual, even if he stands alone in its indulgence. But when it is an almost universal habit, coming down from generation to generation, throwing its creepers and clingers around the social customs and industrial economies of a great nation, it is almost like re-creating a world to change that second nature thus strengthened. This change is slowly working its way in Great Britain—slowly, but perceptibly here and there—thanks to the faithful and persevering efforts put forth by good and true men, to enlighten the subjects of this impoverishing and demoralising custom, which has ruled with such despotism over the laborers of the land. Little by little the proper balance between the Four Great Powers of human necessity,—Food, Drink, Raiment and Housing, so long disturbed by this habit, is being restored. Still, the preponderance of Drink, especially among the agricultural laborers in England, is very striking and sad. As a whole, Beer must still stand before Bread—even before Meat, and before both in many cases, in their expenditures. The man who sat next me, in muddy leggings, and smoking coat, was mildly spoken, quiet, and seemingly thoughtful. He had come for his harvest allowance of 20s. worth of beer. If he abstained from its use on Sundays, he would have a ration of about tenpence's worth daily. That would buy him a large loaf of bread, two good cuts of mutton or beef, and all the potatoes and other vegetables he could eat in a day. But he puts it all into the Jug instead of the Basket. Jug is the juggernaut that crushes his hard earnings in the dust, or, without the figure, distils them into drink. Jug swallows up the first fruits of his industry, and leaves Basket to glean among the sharpest thorns of his poverty. Jug is capricious as well as capacious. It clamors for quality as well as quantity; it is greedy of foaming and beaded liquors. Basket does well if it can bring to the reaper the food of well-kept dogs. In visiting different farms, I have noticed men and women at their luncheons and dinners in the field. A hot mutton chop, or a cut of roast-beef, and a hot potato, seem to be a luxury they never think of in the hardest toil of harvest. Both the meals I have mentioned consist, so far as I have seen, of only two articles of food,—bread and bacon, or bread and cheese. And this bacon is never warm, but laid upon a slice of bread in a thin, cold layer, instead of butter, both being cut down through with a jack-knife into morsels when eaten.

Such is a habit that devours a lion's share of the English laborer's earnings, and leaves Food, Raiment, and Housing to shift for themselves. If he works by the piece and finds his own beer, it costs him more than he pays for house rent, or for bread, or meat, or for clothes for himself and family. If his employer furnishes it or pays him commutation money, it amounts for all his men to a tax of half-a-crown to the acre for his whole farm. There is no earthly reason why agricultural laborers in this country should spend more in drink than those of New England. I am confident that if a census were taken of all the "hired men" of our six states, and a fair average struck, the daily expenditure for drinks would not exceed twopence, or four cents per head, while their average wages would amount to 4s., or 96 cents, per day through the year. Yet our Summers are far hotter and dryer than in England, our labor equally hard, and there is really more natural occasion for drinks in our harvest fields than here. It would require a severe apprenticeship for our men to acquire a taste for sharp ale or strong beer as a beverage under our July sun. A pail or jug of sweetened water, perhaps with a few drops of cider to the pint, to sour it slightly, and a spoonful of ginger stirred in, is our substitute for malt liquor. Sometimes beer made of nothing but hops, water, and a little molasses, is brought into the field, and makes even an exhilarating drink, without any alcoholic effect. Cold coffee, diluted with water, and re-sweetened, is a healthful and grateful luxury to our farm laborers.

It would be a blessed thing for all the outdoor and indoor laborers in this country, if the broad chasm between the strong beer of Old England and the small beer of New England could be bridged, and they be carried across to the shore of a better habit. The farm hands here need a good deal of gentle leading and suggestion in this matter. If some humane and ingenious man would get up a new, cheap, cold drink, which should be nutritious, palatable and exhilarating, without any inebriating property, it would be a boon of immeasurable value. Malt liquors are made in such rivers here, or rather in such lakes with river outlets; there is such a system for their distribution and circulation through every town, village, and hamlet; and they are so temptingly and conveniently kegged, bottled, and jugged, and so handy to be carried out into the field, that the habit of drinking them is almost forced upon the poor man's lips. If a cheaper drink, refreshing and strengthening, could be made equally convenient and attractive, it would greatly help to break this hereditary thraldom to the Beer-Barrel. Another powerful auxiliary to this good work might be contributed in the form of a simple contrivance, which any man of mechanical genius and a kind heart might elaborate. In this go-ahead age, scores of things are made portable that once were fast-anchored solidities. We have portable houses, portable beds, portable stoves and cooking ranges, as well as portable steam-engines. Now, if some benevolent and ingenious man would get up a little portable affair, at the cost of two or three shillings, especially for agricultural laborers in this country, which they could carry with one hand into the field, and by which they could make and keep hot a pot of coffee, cocoa, chocolate, broth or porridge, and also bake a piece of meat and a few potatoes, it would be a real benefaction to thousands, and help them up to the high road of a better condition.

What is the best condition to which the agricultural laborers in Great Britain may ever expect to attain, or to which they may be raised by that benevolent effort now put forth for their elevation? They may all be taught to read and write and do a little in the first three rules of arithmetic. That will raise them to a new status and condition. Education of the masses has become such a vigorous idea with the Government and people of England; so much is doing to make the children of the manufacturing districts pass through the school-room into the factory, carrying with them the ability and taste for reading; ragged schools, working-men's clubs, and institutions for all kinds of cheap learning and gratuitous teaching are multiplying so rapidly; the press is turning out such a world of literature for the homes of the poor, and the English Post, like a beneficent Providence, is distilling such a morning dew of manuscript and printed thoughts over the whole length and breadth of the country, and all these streams of elevating influence are now so tending towards the agricultural laborers, that there is good reason to believe the next generation of them will stand head and shoulders above any preceding one in the stature of intelligence and self-respect. This in itself will give them a new status in society, as beneficial to their employers as to themselves. It will increase their mutual respect, and create a better footing for their relationships.

But the first improvement demanded in their condition, and the most pressingly urgent, is a more comfortable, decent and healthy housing. Until this is effected, all other efforts to raise them mentally and morally must fail of their expected result. The *London Times*, and other metropolitan, and many local, journals publish almost daily distressing accounts of the miserable tenements occupied by the men and women whose labor makes England the garden of fertility and beauty that it is. Editors are making the subject the theme of able and stirring articles, and some of the most eloquent members of Parliament are speaking of it with great power. It is not only generous but just to take the language in which the writers and orators of a country denounce the evils existing in it *cum grano salis*, or with considerable allowance for exaggeration. Their statements and denunciations should not be used against their country as a reproach by the people of another, because they prove an earnest desire and effort to reform abuses which grew up in an unenlightened past. As a specimen of the language which is sometimes held on this subject, I subjoin the following paragraph from the *Saturday Review*, perhaps the most cynical or unsentimental journal in England:—

"There is a wailing for the dirt and vice and misery which must prevail in houses where seven or eight persons, of both sexes and all ages, are penned up together for the night in the one rickety, foul, vermin-hunted bed-room. The picture of agricultural life unrolls itself before us as it is painted by those who know it best. We see the dull, clouded mind, the bovine gaze, the brutality and recklessness, and the simple audacity, and the confessed hatred of his betters, which mark the English peasant, unless some happy fortune has saved him from the general lot, and persuaded him that life has something besides beer that the poor man may have and may relish."

Now this is a sad picture truly. The pen is sharp and cuts like a knife,—but it is the surgeon's knife, not the poisoned barb of a foreigner's taunt. This is the hopeful and promising aspect of these delineations and denunciations of the laboring man's condition. That low, damp, ill-ventilated, contracted room in which he pens his family at night, was, quite likely, constructed in the days of Good Queen Bess, or when "George the Third was King," at the latest. And houses were built for good, substantial farmers in those days which they would hardly house their horses in now. There are hundreds of mechanics and day-laborers in Edinburgh who pen their families nightly in apartments once owned and occupied by Scotch dukes and earls, but which a journeyman shoemaker of New England would be loth to live in rent free. Even the favorite room of Queen Mary, in Holyrood Palace, in which she was wont to tea and talk with Rizzio, would be too small and dim for the shop-parlor of a small London tradesman of the present day. Thus, after all, the low-jointed, low-floored, small-windowed, ill-ventilated cottages now occupied by the agricultural laborers of England were proportionately as good as the houses built at the same period for the farmers of the country, many of which are occupied by farmers now, and the like of which never could be erected again on this island. Indeed, one wonders at finding so many of these old farm houses still inhabited by well-to-do people, who could well afford to live in better buildings.

This, then, is a hopeful sign, and both pledge and proof of progress—that the very cottages of laboring men in England that once figured so poetically in the histories and pictures of rural life, are now being turned inside out to the scrutiny of a more enlightened and benevolent age, revealing conditions that stir up the whole community to painful sensibility and to vigorous efforts to improve them. These cottages were just as low, damp, small and dirty thirty years ago as they are now, and the families "penned" in them at night were doubtless as large, and perhaps more ignorant than those which inhabit them at the present time. It is not the real difference between the actual conditions of the two periods but the difference in the dispositions and perceptions of the public mind, that has produced these humane sensibilities and efforts for the elevation of the ploughers, sowers, reapers and mowers who enrich and beautify this favored land with their patient and poorly-paid labor. And there is no doubt that these newly-awakened sentiments and benevolent activities will carry the day; replacing the present tenements of the agricultural laborers with comfortable, well-built cottages, fitted for the homes of intelligent and virtuous families. This work has commenced in different sections under favorable auspices. Buildings have been erected on an estate here and there which will be likely to serve as models for whole hamlets of new tenements. From what I have heard, I should think that Lord Overstone, of the great banking house of the Lloyds, has produced the best models for cottage homes, on his estates in Northamptonshire. Although built after the most modern and improved plan, and capacious enough to accommodate a considerable family very comfortably, almost elegantly, the yearly rent is only £3, or less than *fifteen dollars*!

Now with a three-pound cottage, having a parlor, kitchen, bed-room and buttery on the lower floor, and an equal number of apartments on the upper; with a forty-rod garden to grow his vegetables, and with a free school for his children at easy walking distance, the agricultural laborer in England will be placed as far forward on the road of improvement as the Government or people, or both, can set him. The rest of the way upward and onward he must make by his own industry, virtue and economy. From this point he must work out his own progress and elevation. No Government, nor any benevolent association, nor general nor private benevolence, can regulate the rate of his wages. The labor market will determine that, just as the Corn Exchange does the price of wheat. But there is one thing he can do to raise himself in civil stature, moral growth, and domestic comfort. He may empty the Jug into the Basket. He and his family may consume in solids what they now do in frothy fluids. They may exchange their scanty dinner of cold bacon and bread for one of roast beef and plum pudding, by substituting cold coffee, cocoa or pure water for strong beer. Or, if they are content to go on with their old fare of food, they may save the money they expended in ale for the rent of one or two acres of land, for a cow, or for two or three pigs, or deposit it weekly in the Post-Office Savings' Bank, until it shall amount to a sum sufficient to enable them to set up a little independent business of their own.

Here, then, are three great steps indispensable for the elevation of the agricultural laborers of Great Britain to the highest level in society which they can reach and maintain. Two of these the Government, or the land-owners, or both, must take. They are Improved Dwellings and Free and Accessible Education. These the laborer cannot provide for himself and family. It is utterly beyond his ability to do it. The third, last, long step must depend entirely upon himself; though he may be helped on by sympathy, suggestion, and encouragement from those who know how hard a thing it is for the fixed appetites to break through the meshes of habit. He must make drink the cheapest of human necessities. He must exchange Beer for Bread, for clothes, for books, or for things that give permanent comfort and enjoyment. When these three steps are accomplished, the British laborer will stand before his country in the best position it can give him. And I believe it will be a position which will make him contented and happy, and be satisfactory to all classes of the people.

After all that can be done for them, the wages of the agricultural laborers of Great Britain cannot be expected to exceed, on an average, twelve shillings a week, or about half the price of the same labor in America. Their rent and clothes cost them, perhaps, less than half the sum paid by our farm hands for the same items of expenditure. Their food must also cost only about half of what our men pay, who would think they were poor indeed if they could not have hot meat breakfasts, roast or boiled beef dinners and cold meat suppers, with the usual sprinkling of puddings, pies, and cakes, and tea sweetened with loaf sugar. Thus, after all, put the English laborer in the position suggested; give him such a three-pound cottage and garden as Lord Overstone provides; give his children free and convenient schooling; then let him exchange his ale for nutritious and almost costless drinks, and if he is still able to live for a few years on his old food-fare, he may work his way up to a very comfortable condition with his twelve shillings a week, besides his beer-money. On these conditions he would be able almost to run neck and neck with our hired men in the matter of saving money "for a rainy day," or for raising himself to a higher position.

We will put them side by side, after the suggested improvements have been realised; assuming each has a wife, with two children too young to earn anything at field work.

American Laborer at 24s per week
English Laborer, at 12s per week

Weekly Expense for:--	$ c.	s. d
Food	3 50 =	14 7
Rent and Taxes	0 67 =	2 9
Fuel, average of the year	O 48 =	2 O
For Clothes	1 0 =	4 2
Total Weekly Expenses	5 65 =	23 6

Weekly Expense for:--	s. d.	$ c
Food	7 3 =	1 75
Rent	1 2 =	O 28
Fuel	1 O =	O 24
Clothes	2 1 =	0 50
Total Weekly Expenses	11 6 =	2 77

I think the American reader, who is personally acquainted with the habits and domestic economy of our farm laborers, will regard this estimate of their expenditures as quite moderate. I have assumed, in both cases, that no time is lost in the week on account of sickness, or of weather, or lack of employment; and all the incidental expenses I have included in the four general items given. It must also be conceded that our farm hands do not average more than twenty-four English shillings, or $5 75c., per week, through all the seasons of the year. The amount of expenditure allowed in the foregoing estimate enables them to support themselves and their families comfortably, if they are temperate and industrious; to clothe and educate their children; to make bright and pleasant homes, with well-spread tables, and to have respectable seats in church on the Sabbath. On the other hand, we have assigned to the English agricultural laborer what he would regard a proportionately comfortable allowance for the wants of a week. We may not have divided it correctly, but the total of the items is as great as he would expect to expend on the current necessities of seven days. I doubt if one in a thousand of the farm laborers of Great Britain lays out more than the sum we have allotted for one week's food, rent, and fuel and clothes. We then reach this result of the balance-sheet of the two men. Their weekly savings hardly differ by a penny; each amounting to about 5d., or 10 cents. At first sight, it might seem, from this result, that the English farm laborer earns half as much, lives half as well, and saves as much as the American. But he has a resource for increasing his weekly savings which his American competitor would work his fingers to the bone before he would employ. His wife is able and willing to go with him into the field and earn from three to five shillings a week. Then, if he commutes with his employer, he will receive from him 4d. daily, or 2s. a week, for beer-money. Thus, if he and his wife are willing to live, as such families do now, on bread, bacon and cheese, and such vegetables as they can grow in their garden, they may lay up, from their joint earnings, a dollar, or four shillings a week, provided a sufficiently stimulating object be set before them. To me it is surprising that they sustain so much human life on such small means. They are often reproached for their want of wise economy; but never was more keen ingenuity, more close balancing of pennies against provisions than a great many of them practice and teach. Let the most astute or utilitarian of social economists try the experiment of housing, feeding and clothing himself, wife and six children too young to earn anything, on ten or twelve shillings a week; and he will learn something that his philosophy never dreamed of.

Even while bending under the weight of the beer-barrel, thousands of agricultural laborers in England have accomplished wonders by their indefatigable industry, integrity and economy. Put a future before them with a sun in it—some object they may reach that is worth a life's effort, and as large a proportion of them will work for it as you will find in any other country. A servant girl told me recently that her father was a Devonshire laborer, who worked the best years of his life for seven shillings a week, and her mother for three, when they had half a dozen children to feed and clothe. Yet, by that unflagging industry and ingenious economy with which thousands wrestle with the necessities of such a life and throw them, too, they put saving to saving, until they were able to rent an acre of orcharding, a large garden for vegetables, then buy a donkey and cart, then a pony and cart, and load and drive them both to market with their own and their neighbors' produce, starting from home at two in the morning. In a few years they were able to open a little grocery and provision shop, and are now taking their rank among the tradespeople of the village. But if the farm servants of England could only be induced to give up beer and lay by the money paid them as a substitute, it alone would raise them to a new condition of comfort, even independence. At 4d. a day commutation money, they would have each £5 at the end of the year. That would pay the rent of two acres of land here; or it would buy five on the Illinois Central Railroad. Three years' beer-money would pay for those rich prairie acres, his fare by sea and land to them, and leave him £3 in his pocket to begin their cultivation with. Three years of this saving would make almost a new man of him at home, in the way of self-respect, comfort and progress. It would be a "nest-egg," to which hope, habit and a strengthening ambition would add others of larger size and value from year to year.

Give, then, the British agricultural laborer good, healthy Housing, Free Schooling, and let him empty the Jug into the Basket, and he may work his way up to a very comfortable condition at home. But if he should prefer to go to Australia or America, where land is cheap and labor dear, in a few years he may save enough to take him to either continent, with sufficient left in his pocket to begin life in a new world.

CHAPTER XII.

FARM GAME—HALLETT WHEAT—OUNDLE—COUNTRY BRIDGES—FOTHERINGAY CASTLE—QUEEN MARY'S IMPRISONMENT AND EXECUTION—BURGHLEY HOUSE: THE PARK, AVENUES, ELMS, AND OAKS—THOUGHTS ON TREES, ENGLISH AND AMERICAN.

Having now pursued a westerly direction until I was in the range of a continuous upland section of country, I took a northward course and walked on to Oundle, a goodly town in Northamptonshire, as unique as its name. On the way, in crossing over to another turnpike road, I passed through a large tract of land in a very *deshabille* condition, rough, boggy and bushy. I soon found it was a game-growing estate, and very productive of all sorts of birds and small quadrupeds. The fields I crossed showed a promising crop of hares and rabbits; and doubtless there were more partridges on that square mile than in the whole State of Connecticut. This is a characteristic of the country which will strike an American, at his first visit, with wonder. He will see hares and rabbits bobbing about on common farms, and partridges in broods, like separate flocks of hens and chickens, in fields of grain, within a stone's throw of the farmer's house. I doubt if any county in New England produces so many in a year as the holding of Mr. Samuel Jonas already described. Rabbits have been put out of the pale of protection somewhat recently, I believe, and branded with the bad name of *vermin*; so that the tenant farmer may kill them on his occupation without leave or license from the landlord. It may indicate their number to state the fact, that one hundred and twenty-five head of them were killed in one day's shooting on Mr. Jonas's estate by his sons and some of their friends.

It was market day in Oundle, and I had the pleasure of sitting down to dinner with a large company of farmers and cattle and corn-dealers. They were intelligent, substantial-looking men, with no occupational peculiarity of dress or language to distinguish them from ordinary middle-class gentlemen engaged in trade or manufacture. Indeed, the old-fashioned English farmer, of the great, round, purply-red face, aldermanic stature, and costume of fifty years ago, speaking the dialect of his county with such inimitable accent, is fast going out. I have not seen one during my present sojourn in England. I fear he has disappeared altogether with the old stage-coach, and that we have not pictures enough of him left to give the rising generation any correct notion of what he was, and how he looked. It may be a proper and utilitarian change, but one can hardly notice without regret what transformations the railway *regime* has wrought in customs and habits which once individualised a country and people. A kind of French centralisation in the world of fashion has been established, which has over-ridden and obliterated all the dress boundaries of civilised nations. All the upper and middle classes of Christendom centre themselves to one focus of taste and merge into one plastic commonwealth, to be shaped and moulded virtually by a common tailor. Their coats, vests, pantaloons, boots and shoes are made substantially after the same pattern. For a while, hats stood out with some show of pluck and patriotism, and made a stand for national individuality, but it was in vain. They, too, succumbed to the inexorable law of Uniformity. That law was liberal in one respect. It did not insist that the stove-pipe form should rule inflexibly. It admitted several variations, including wide-awakes, pliable felts, and that little, squat, lackadaisical, round-crown, narrow-brimmed thing worn by the Prince of Wales in the photographs taken of him and the Princess at Sandringham. But this has come to be the rule: that hats shall no longer represent distinct nationalities; that they shall be interchangeable in all civilised communities; in a word, that neither Englishman, American, French nor German shall be known by his hat, whatever be the form or material of its body or brim. If there were a southern county in England where the mercury stood at 100 degrees in the shade for two or three summer months, the upper classes in it would don, without any hesitation, the wide, flappy broadbrims of California, and still be in the fashion,—that is, variety in uniformity. The peasantry, or the lowest laboring classes of European countries, are now, and will remain perhaps for a century to come, the only conservators of the distinctive national costumes of bygone generations.

During the conversation at the table, a farmer exhibited a head of the Hallett wheat, which he had grown on his land. I never saw anything to equal it, in any country in which I have travelled. It was nearly six inches in length, and seeded large and plump from top to bottom. This is a variety produced by Mr. Hallett, of Brighton, and is creating no little interest among English grain-growers. Lord Burghley, who had tested its properties, thus describes it, in a speech before the Northamptonshire Agricultural Society last summer:—

"At the Battersea Show last year, my attention was called to some enormous ears of wheat, which I thought could not have been grown in England. For, although the British farmer can grow corn with anyone, I had never seen such wheat here, and thought it must be foreign wheat. I went to the person who was threshing some out, and having been informed that it was sown only with one seed in a hole, I procured some of Mr. Hallett, of Brighton; and, being anxious to try the system, I planted it according to Mr. Hallett's directions—one grain in a hole, the holes nine and a half inches apart, with six inches between the rows. To satisfy myself on the subject, I also planted some according to Stephen's instructions, who said three grains in a hole would produce the most profitable return. I also planted some two grains in a hole. I sowed the grain at the end of last September, on bad land, over an old quarry, and except some stiff clay at the bottom of it, there was nothing in it good for wheat. The other day I counted the stalks of all three. On Mr. Stephen's plan of three grains in a hole, there were eighteen stalks; with two grains in a hole, there was about the same number; but with one seed in a hole, the lowest number of stalks was sixteen, and the highest twenty-two. I planted only about half an acre as a trial, and when I left home a few days since, it looked as much like eight quarters (sixty-four bushels) to the acre as any I have seen. The ears are something enormous. I would certainly recommend every farmer to make his own experiments, for if it succeeds, it will prove a great economy of seed; and drills to distribute it fairly are to be had."

Truly one of Hallett's wheat ears might displace the old *cornucopia* in that picture of happy abundance so familiar to old and young. Here are twenty ears from one seed, containing probably a thousand grains. The increase of a thousand-fold, or half that ratio, is prodigious, having nothing to equal it in the vegetable world that we know of. If one bushel of seed wheat could be so distributed by a drill as to produce 500 or 250 bushels at the harvest, certainly

the staff of life would be greatly cheapened to the millions who lean upon it alone for subsistence.

From Oundle I walked the next day to Stamford, a good, solid, old English town, sitting on the corners of three counties, and on three layers of history, Saxon, Dane and Norman. The first object of interest was a stone bridge over the Nen at Oundle. It is a grand structure to span such a little river. It must have cost three times as much as "The Great Bridge" over the Connecticut at Hartford; and yet the stream it crosses is a mere rivulet compared with our New England river. "The bridge with wooden piers" is a fabric of fancy to most English people. They have read of such a thing in Longfellow's poems, but hardly realise that it exists still in civilised countries. Here bridges are works of art as well as of utility, and rank next to the grand old cathedrals and parish churches for solidity and symmetry. Their stone arches are frequently turned with a grace as fine as any in St. Paul's, and their balustrades and butments often approach the domain of sculpture.

Crossing the Nen, I followed it for several miles in a northerly direction. I soon came to a rather low, level section of the road, and noticed stones placed at the side of it, at narrow intervals, for a long distance to the very foot of a village situated on a rising ground. These stones were evidently taken from some ancient edifice, for many of them bore the marks of the old cathedral or castle chisel. They were the foot-tracks of a ruined monument of dark and painful history. More than this might be said of them. They were the blood-drops of a monstrosity chased from its den and hunted down by the people, who shuddered with horror at its sanguinary record of violence and wrong. As I approached the quiet village, whose pleasant-faced houses, great and small, looked like a congregation of old and young sitting reverently around the parish church and listening to the preaching of the belfry, I saw where these stones came from. There, on that green, ridgy slope, where the lambs lay in the sun by the river, these stones, and a million more scattered hither and thither, once stood in walls high, hideous and wrathful, for half a dozen centuries and more. If the breathings of human woe, if the midnight misery of wretched, broken hearts, could have penetrated these stones, one might almost fancy that they would have sweat with human histories in the ditch where they lay, and discolored the puddles they bridged with the bitter distilment of grief centuries old. On that gentle rising from the little Nen stood Fotheringay Castle. That central depression among the soft-carpeted ridges marks the site of the *donjon* huge and horrid, where many a knight and lady of noble blood was pinioned or penned in darkness and hopeless duress centuries before the unfortunate Mary was born. There nearly half the sad years of her young life and beauty were prisoned. There she pined in the sickness of hope deferred, in the corroding anguish of dread uncertainty, for a space as wide as that between the baptismal font and presentation at Elizabeth's court. There she laid her white neck upon the block. There fell the broad axe of Elizabeth's envy, fear and hate. There fell the fair-haired head that once gilded a crown and wore all the glory of regal courts—still beautiful in the setting light of farewell thoughts.

It may be truly said of Fotheringay Castle, that not one stone is left upon another to mark its foundations. Not Fleet-street Prison, nor the Bastille itself, went out under a heavier weight of popular odium. Although public sentiment, as well as the personal taste and interest of their proprietors, has favored the preservation of the ruins of old castles and abbeys in Great Britain, Fotheringay bore, branded deep in its forehead, the mark of Cain, and every man's hand, of the last generation, seemed to have been turned against it. It has not only been demolished, but the *debris* have been scattered far and wide, and devoted to uses which they scarcely honor. You will see the well-faced stones for miles around, in garden walls, pavements, cottage hearths and chimneys, in stables and cow-houses.

In Oundle, the principal hotel, a large castellated building, shows its whole front built of them.

The great lion of Stamford is the Burghley House, the palace of the Marquis of Exeter. It may be called so without exaggeration of its magnificence as a building or of the extent and grandeur of its surroundings. The edifice itself would cut up into nearly half a dozen "White Houses," such as we install our American Presidents in at Washington. Certainly, in any point of view, it is large and splendid enough for the residence of an emperor and his *suite*. Its towers, turrets and spires present a picturesque grove of architecture of different ages, and its windows, it is said, equal in number all the days of the year. It was not open to the public the day I was in Stamford, so I could only walk around it and estimate its interior by its external grandeur.

But there was an outside world of architecture in the park of sublimer features to me than even the great palace itself, with all its ornate and elaborate sculpture. It was the architecture of the majestic elms and oaks that stood in long ranks and folded their hands, high up in the blue sky, above the finely-gravelled walks that radiated outward in different directions. They all wore the angles and arches of the Gothic order and the imperial belt of several centuries. I walked down one long avenue and counted them on either side. There were not sixty on both; yet their green and graceful roofage reached a full third of a mile. Not sixty to pillar and turn such an arch as that! I sat down on a seat at the end to think of it. There was a morning service going on in this Cathedral of Nature. The dew-moistened, foliated arches so lofty, so interwebbed with wavy, waky spangles of sky, were all set to the music of the anthem. "The street musicians of the heavenly city" were singing one of its happiest hymns out of their mellow throats. The long and lofty orchestra was full of them. Their twittering treble shook the leaves with its breath, as it filtered down and flooded the temple below. Beautiful is this building of God! Beautiful and blessed are these morning singing-birds of His praise! Amen!

But do not go yet. No; I will not. Here is the only book I carry with me on this walk—a Hebrew Psalter, stowed away in my knapsack. I will open it here and now, and the first words my eye lights upon shall be a text for a few thoughts on this scene and scenery. And here they are,—seemingly not apposite to this line of reflection, yet running parallel to it very closely:

[HEBREW PHRASE]

The best English that can be given of these words we have in our translation: "Blessed is he who, passing through the valley of Baca, maketh it a well." Why so? On what ground? If a man had settled down in that valley for life, there would have been no merit in his making it a well. It might, in that case, have been an act of lean-hearted selfishness on his part. Further than this, a man might have done it who could have had the heart to wall it in from the reach of thirsty travellers. No such man was meant in the blessing; nor any man resident in or near the valley. It was he who was "passing through" it, and who stopped, not to search for a dribbling vein of water to satisfy his own momentary thirst, but to make a well, broad and deep, after the oriental circumference, at which all future travellers that way might drink with gladness. That was the man on whom the blessing rested as a *condition*, not as a *wish*. Look at the word, and get the right meaning of it. It is [HEBREW WORD], not [HEBREW WORD]; it is a blessedness, not a benediction. It means a permanent reality of happiness, like that of Obededom, not a cheap "I thank you!" or "the Lord bless you!" from here and there a man or woman who appreciates the benefaction.

And he deserves the same who, "passing through" the short years of man's life here on earth, plants trees like the living, lofty columns of this long cathedral aisle. How unselfish and generous is this gift to coming generations! How inestimable in its value and surpassing the worth of wealth!—surpassing the measurement of gold and silver! From my seat here, I look up to the magnificent frontage of that baronial palace. I see its towers, turrets and minarets; its grand and sculptured gateways and portals through this long, leaf-arched aisle. Not forty, but nearer four hundred years, doubtless, was that pile in building. Architecture of the pre-Norman period, and of all subsequent or cognate orders, diversifies the tastes and shapings of the structure. Suppose the whole should take fire to-night and burn to the ground. The wealth of the owner could command genius, skill and labor enough to rebuild it in three years, perhaps in one. The Czar of all the Russias did as large a thing once as this last, in the reconstruction of a palace. Perhaps the building is insured for its positive value, and the insurance money would erect a better one. But lift an axe upon that tall centurion of these templed elms. Cut through the closely-grained rings that register each succeeding year of two centuries. Hear the peculiar sounding of the heart-strokes, when the lofty, well-poised structure is balancing itself, and quivering through every fibre and leaf and twig on the few unsevered tendons that have not yet felt the keen edge of the woodman's steel. See the first leaning it cannot recover. Hear the first cracking of the central vertebra; then the mournful, moaning whir in the air; then the tremendous crash upon the green earth; the vibration of the mighty trunk on the ground, like the writhing and tremor of an ox struck by the butcher's axe; the rebound into the air of dismembered branches; the frightened flight of leaves and dust, and all the other distractions of that hour of death and destruction. Look upon that ruin! The wealth, genius and labor that could build a hundred Windsor Castles, and rebuild all the cathedrals of England in a decade, could not rebuild in two centuries that elm to the life and stature you levelled to the dust in two hours.

Put, then, the man who plants trees for posterity with him who, "passing through the valley of Baca, maketh it a well." Put him under the same blessing of his kind, for he deserves it. He gives them the richest earthly gift that a man can give to a coming generation. In a practical sense, he gives them *time*. He gives them a whole century, as an extra. If they would pay a gold sovereign for every solid inch of oak, they could not hire one built to the stature of one of these trees in less than two centuries' time, though they dug about it and nursed it as the man did the vine in Scripture. Blessed be the builders of these living temples of Nature! Blessed be the man, rich or poor, old or young, especially the old, who sets his heart and hand to this cheap but sublime and priceless architecture.

Let connoisseurs who have seen Memphis, Nineveh, Athens, Rome, or any or all of the great cities of the East, ancient or modern, come and sit here, and look at this lofty corridor, and mark the orders and graces of its architecture. What did the Ptolemies, their predecessors or successors in Egypt, or sovereigns of Chaldaic names, in Assyria, or ambitious builders in the ages of Pericles or Augustus, in Greece or Rome? Their structures were the wonders of the world. Mighty men they were, whose will was law, whose subjects worked it out to its wildest impulse without a murmur or a reward. But who built this sixty-columned temple, and bent these lofty arches? Two or three centuries ago, two men in coarse garb, and, it may be, in wooden shoes, came here with a donkey, bearing on its back a bundle of little elms, each of a finger's girth. They came with the rude pick and spade of that time; and, in the first six working hours of the day, they dug thirty holes on this side of the aisle, and planted in them half the tiny trees of their bundle. They then sat down at noon to their bread and cheese and, most likely, a mug of ale, and talked of small, home matters, just as if they were dibbling in a small patch of wheat or potatoes. They then went to work again and planted the other row; and, as the sun was going down, they straightened their backs, and, with hands stayed upon their hips, looked up and down the two lines and thought they would pass muster and please the master. Then they shouldered their brightened tools and went home to their low, dark cottages, discussing the prices of bread, beer and bacon, and whether the likes of them could manage to keep a pig and make a little meat in the year for themselves.

That is the story of this most magnificent structure to which you look up with such admiration. Those two men in smock frocks, each with a pocket full of bread and cheese, were the Michael Angelos of this lofty St. Peter's. That donkey, with its worn panniers, was the only witness and helper of their work. And it was the work of a day! They may have been paid two English shillings for it. The little trees may have cost two shillings more, if taken from another estate. The donkey's day was worth sixpence. O, wooden-shoed Ptolemies! what a day's work was that for the world! They thought nothing of it— nothing more than they would of transplanting sixty cabbages. They most likely did the same thing the next day, and for most of the days of that year, and of the next year, until all these undulating acres were planted with trees of every kind that could grow in these latitudes. How cheap, but priceless, is the gift of such trees to mankind! What a wealth, what a glory of them can even a poor, laboring man give to a coming generation! They are the most generous crops ever sown by human hands. All others the sower reaps and garners into his own personal enjoyment; but this yields its best harvest to those who come after him. This is a seeding for posterity. From this well of Baca shall they draw the cooling luxury of the gift when the hands that made it shall have gone to dust.

And this is a good place and time to think of home—of what we begin to hear called by her younger children, *Old* New England. Trees with us have passed through the two periods specified by Solomon—"a time to plant and a time to pluck up." The last came first and lasted for a century. Trees were the natural enemies to the first settlers, and ranked in their estimation with the wild Indians, wolves and bears. It was their first, great business to cut them down, both great and small. Forests fell before the woodman's axe. It made clean work, and seldom spared an oak or an elm. But, at the end of a century, the people relented and felt their mistake. Then commenced "the time to plant;" first in and around cities like Boston, Hartford, and New Haven, then about villages and private homesteads. Tree-planting for use and ornament marks and measures the footsteps of our civilization. The present generation is reaping a full reward of this gift to the next. Every village now is coming to be embowered in this green legacy to the future; like a young mother decorating a Christmas-tree for her children. Towns two hundred years old are taking the names of this diversified architecture, and they glory in the title. New Haven, with a college second to none on the American Continent, loves to be called "The Elm City," before any other name. This generous and elevating taste is making its way from ocean to ocean, even marking the sites of towns and villages before they are built. I believe there is an act of the Connecticut Legislature now in force, which allows every farmer a certain sum of money for every tree he plants along the public roadside of his fields. The object of this is to line all the highways of the State with ornamental trees, so that each shall be a well-shaded avenue. What a gift to another generation that simple act is intended to make! What a world of wonder and delight will our little State be to European travellers and tourists of the next century, if this measure shall be carried out! If a few miles of such avenues as Burghley Park and Chatsworth present, command such admiration, what sentiments would a continuous avenue of trees of equal size from Hartford to New Haven inspire!

While on this line of reflection, I will mention a case of monumental tree-planting in New England, not very widely known there. A small town, in the heart of Massachusetts, was stirred to the liveliest emotion, with all the rest in her borders, by the Declaration of Independence in 1776. Different communities expressed their sense of the importance of this event in different ways,

most of which were noisy and excited. But the good people of this rural parish came together, and, at a happy suggestion from some one of their number, agreed to spend the day in planting trees to commemorate the momentous transaction. They forthwith set to work, young and old, and planted first a double row on each side of the walk from the main road up "The Green" to their church door; then a row on each side of the public highway passing through the village, for nearly a mile in each direction. There was a blessed day's work for them, their children and children's children. Every hand that wielded a spade, or held up a treelet until its roots were covered with earth, has long since lost its cunning; but the tall, green monuments they erected to the memory of the most momentous day in American history, stand in unbroken ranks, the glory of the village.

Although America will never equal England, probably, in compact and picturesque "plantations," or "woods," covering hundreds of acres, all planted by hand, our shade-trees will outnumber hers, and surpass them in picturesque distribution and arrangement, when our popular programme is fully carried out. In two or three important particulars, we have a considerable advantage over this country in respect to this tasteful embellishment. In the first place, all the farmers in America own the lands they cultivate, and, on an average, two sides of every farm front upon a public road. Two or three days' work suffices for planting a row of trees the whole length of this frontage, or the roadside of the farmer's fence or wall. This is being done more and more extensively from year to year, generally under the influence of public taste and custom, and sometimes under the stimulus of governmental compensation, as in Connecticut. Thus, in the life of the present generation, all our main roads and cross-roads may become arched and shaded avenues, giving the whole landscape of the country an aspect which no other land will present.

Then we have another great advantage which England can never attain un-

til she learns how to consume her coal smoke. Our wood and anthracite fires make no smoke to retard the growth or blacken the foliage of our trees. Thus we may have them in standing armies, tall and green, lining the streets, and overtopping the houses of our largest cities; filtering with their wholesome leafage the air breathed by the people. New Haven and Cleveland are good specimens of beautifully-shaded towns.

There is a third circumstance in our favor as yet, and of no little value. The grand old English oak and elm are magnificent trees, in park or hedge-row here. The horse-chestnut, lime, beech and ash grow to a size that you will not see in America. The Spanish chestnut, a larger and coarser tree than our American, reaches an enormous girth and spread. The pines, larches and firs abound. Then there are tree-hunters exploring all the continents, and bringing new species from Japan and other antipodean countries. But as yet, our maples have never been introduced; and without these the tree-world of any country must ever lack a beautiful feature, both in spring, summer and autumn, especially in the latter. Our autumnal scenery without the maple, would be like the play of Hamlet with Hamlet left out; or like a royal court without a queen. Few Americans, even loudest in its praise, realise how much of the glory of our Indian summer landscape is shed upon it by this single tree. At all the Flower Shows I have seen in England and France, I have never beheld a bouquet so glorious and beautiful as a little islet in a small pellucid lake in Maine, filled to the brim, and rounded up like a full-blown rose, with firs, larches, white birches and soft maples, with a little sprinkling of the sumach. An early frost had touched the group with every tint of the rainbow, and there it stood in the ruddy glow of the Indian summer, looking at its face in the liquid mirror that smiled, still as glass, under its feet.

I was much pleased to notice what honor was put upon one of our humble and despised trees in Burghley House park, as in the grounds of other noblemen. There was not one that spread such delicate and graceful tresses on the breeze as our White Birch; not one that fanned it with such a gentle, musical flutter of silver-lined leaves; not one that wore a bodice of such virgin white from head to foot, or that showed such long, tapering fingers against the sky. I was glad to see such justice done to a tree in the noblest parks in England, which with us has been treated with such disdain and contumely. When I saw it here in such glory and honor, and thought how, notwithstanding its Caucasian complexion, it is regarded as a nuisance in our woods, meadows and pastures, so that any man who owns, or can borrow an axe, may cut it down without leave or license wherever he finds it—when I saw this disparity in its status in the two Englands, I resolved to plead its cause in my own with new zeal and fidelity.

CHAPTER XIII.

WALK TO OAKHAM—THE ENGLISH AND AMERICAN SPRING—THE ENGLISH GENTRY—A SPECIMEN OF THE CLASS—MELTON MOWBRAY AND ITS SPECIALITIES—BELVOIR VALE AND ITS BEAUTY—THOUGHTS ON THE BLIND PAINTER.

From Stamford to Oakham was an afternoon walk which I greatly enjoyed. This was the first week of harvest, and the first of August. How wonderfully the seasons are localised and subdivided. How diversified is the economy of light and heat! That field of wheat, thick, tall and ripe for the sickle, was green and apparently growing through all the months of last winter. What a phenomenon it would have been, on the first of February last, to a New England farmer, suddenly transported from his snow-buried hills to the view of this landscape the same day! Not a spire of grass or grain was alive when he left his own homestead. All was cold and dead. The very earth was frozen to the solid-

ity and sound of granite. It was a relief to his eye to see the snow fall upon the scene and hide it two feet deep for months. He looks upon this, then upon the one he left behind. This looks full of luxuriant life, as green as his in May. It has three months' start of his dead and buried crop. He walks across it; his shoes sink almost to the instep in the soft soil. He sees birds hopping about in it without overcoats. Surely, he says to himself, this is a favored land. Here it lies on the latitudes of Labrador, and yet its midwinter fields are as green as ours in the last month of Spring. At this rate the farmers here must harvest their wheat before the ears of mine are formed. But he counts without Nature. The American sun overtakes and distances the English by a full month. Here is the compensation for six consecutive months in which the New England farmer must house his plough and not turn a furrow.

Doubtless, as much light and heat brighten and warm one country as the other in the aggregate of a year. But there is a great difference in the economy of distribution. In England, the sun spreads its warmth more evenly over the four seasons of the year. What it withholds from Summer it gives to Winter, and makes it wear the face of Spring through its shortest and coldest days. But then Spring loses a little from this equalising dispensation. It is not the resurrection from death and the grave as it is in America. Children are not waiting here at the sepulchre of the season, as with us, watching and listening for its little *Bluebird* angel to warble from the first budding tree top, "*It is risen!*" They do not come running home with happy eyes, dancing for joy, and shouting through the half open door, "O, mother, Spring has come! We've heard the *Bluebird*! Hurrah! Spring has come. We saw the *Phebee* on the top of the sawmill!" Here Spring makes no sensation; takes no sudden leap into the seat of Winter, but comes in gently, like the law of primogeniture or the British Constitution. It is slow and decorous in its movements. It is conservative, treats its predecessor with much deference, and

makes no sudden and radical changes in the face of things. It comes in with no Lord Mayor's Day, and blows no trumpets, and bends no triumphal arches to grace its *entree*. Few new voices in the tree-tops hail its advent. No choirs of tree-toads fiddle in the fens. No congregation of frogs at twilight gather to the green edges of the unfettered pond to sing their Old Hundred, led by venerable Signor Cronker, in his bright, buskin doublet, mounted on a floating stump, and beating time with a bulrush. No *Shad-spirits* with invisible wings, perform their undulating vespers in the heavens, to let the fishermen know that it is time to look to their nets. Even the hens of the farm-yard cackle with no new tone of hope and animation at the birth of the English Spring. The fact is, it is a baby three months old when it is baptised. It is really born at Christmas instead of Easter, and makes no more stir in the family circle of the seasons than any familiar face would at a farmer's table.

In a utilitarian point of view, it is certainly an immense advantage to all classes in this country, that Nature has tempered her climates to it in this kindly way. I will not run off upon that line of reflection here, but will make it the subject of a few thoughts somewhere this side of John O'Groat's. But what England gains over us in the practical, she loses in the poetical, in this economy of the seasons. Her Spring does not thrill like a sudden revelation, as with us. It does not come out like the new moon, hanging its delicate silver crescent in the western pathway of the setting sun, which everybody tries to see first over the right shoulder, for the very luck of the coincidence. Still, both countries should be contented and happy under this dispensation of Nature. The balance is very satisfactory, and well suited to the character and habits of the two peoples. The Americans are more radical and sensational than the English; more given to sudden changes and stirring events. Sterne generally gets the credit of saying that pretty thought first, "Providence tempers the wind to the shorn lamb." A French writer puts it the other way, and more practically: "Providence tempers the wool of the lamb to the wind." This is far better and more natural. But it may be truly said that Providence tempers the seasons to the temperaments and customs of the two nations.

Just before reaching Oakham, I passed a grand mansion, standing far back from the turnpike road, on a commanding eminence, flanked with extensive plantations. The wide avenue leading to it looked a full mile in length. Lawns and lakes, which mirrored the trees with equal distinctness, suffused the landscape of the park like evening smiles of Nature. It was indeed a goodly heritage for one man; and he only mounted a plain *Mr.* to his name, although I learned that he could count his farms by the dozen. I was told that the annual dinner given to his tenant farmers came off the previous day at the inn where I lodged. A sumptuous banquet was provided for them, presided over by the steward of the estate; as the great *Mr.* did not honor the plebeian company with his presence. This is a feature of the structure of English society which the best read American would not be likely to recognise without travelling somewhat extensively in the country. The British Nobility, the great, world-renowned Middle Class, and the poor laboring population, constitute the three great divisions of the people and include them all in his mind. He is apt to leave out of count the Gentry, the great untitled MISTERS, who come in between the nobility and middle-men, and constitute the connecting link between them. "The fine old English gentleman, all of the olden time," is supposed to belong to this class. They make up most of "the old county families," of which you hear more than you read. They are generally large landholders, owning from twenty to one hundred farms. They live in grand old mansions, surrounded with liveried servants, and inspire a mild awe and respectful admiration, not only in the common country people, but in the minds of persons in whom an American would not look for such homage to untitled rank. They hunt

with horses and dogs over the grounds of their tenant farmers, and the latter often act as game-beaters for them at their "shootings." When one of them owns a whole village, church and all, he is generally called "the Squire," but most of them are squired without the definite article. They still boast of as good specimens of "the fine old English gentleman" as the country can show; and I am inclined to think it is not an unfounded pretension, although I have not yet come in contact with many of the class.

One of this county squirocracy I know personally and well,—and other Americans know him as well as myself,—who, though living in a palace of his own, once occupied by an exiled French sovereign, is just as simple and honest as a child in every feature of his disposition and deportment. Every year he has a Festival in his park, lasting two or three days. It is a kind of out-door Parliament and a Greenwich Fair combined, as it would seem at first sight to an incidental spectator. I do not believe anything in the rest of the wide world could equal this gathering, for many peculiar features of enjoyment. It is made up of both sexes and all ages and conditions; especially of the laboring classes. They come out strong on these occasions. The round and red faced boys and girls of villages and hamlets for a great distance around look forward to this annual frolic with exhilarating expectation. Never was romping and racing and the amorous forfeit plays of the ring got up under more favorable auspices, or with more pleasant surroundings. It would do any man's heart good, who was ever a genuine boy, to see the venerable squire and his lady presiding over a race between competing couples of ploughmens' boys, from ten to fifteen years of age, running their rounds in the park, bare-footed, bare-headed, with faces as round and red as a ripe pumpkin, and hair of the same color whipping the air as they neck-and-neck it in the middle of the heat. When the winners of the prizes receive their rewards at his hands, his kind words and the radiant benevolence of his face they value more than the conquest and the

coins they win.

Then there are intellectual entertainments and deliberative proceedings of grave moment arranged for the elder portion of the great congregation. While groups of blushing lads and lasses are hunting the handkerchief in the hustle and tussle of the ring under the great, solemn elms, a scene may be witnessed on the lawn nearer the mansion that ought to have been painted long ago. Two or three double-horse wagons are ranged end to end in the shade, and planks are placed along from one end to the other, making a continuous seat for a score or two of orators. In front of this dozen-wheeled tribune rows of seats, capable of holding several hundred persons, are arranged within hearing distance. When these are filled and surrounded by a standing wall of men and women, three or four deep, and when the orators of the day ascend over the wheels to the long wagon-seat, you have a scene and an assembly the like of which you find nowhere else in Christendom. No Saxon parliament of the Heptarchy could "hold a candle to it. " Never, in any age or country of free speech, did individual ideas, idiosyncrasies, and liberty of conscience have freer scope and play. Never did all the isms of philanthropy, politics, or of social and moral reform generally have such a harmonious trysting time of it. Never was there a platform erected for discussing things local and general so catholic as the one now resting upon the wheels of those farm wagons. Every year the bland and venerable host succeeds in widening the area of debate. I was invited to be present at the Festival this year, but was too far on the road to John O'Groat's to participate in a pleasure I have often enjoyed. But I read his *resume* of the year's doings, aspects and prospects from Japan to Hudson's Bay with lively interest and valuable instruction. He seldom presides himself as chairman, but leaves that post of honor to be filled, if possible, by the citizen of some foreign country, if he can speak English tolerably. This gives a more cosmopolitan aspect to the assembly. But he himself always makes what in

Parliament would be called "a financial statement," without the reference to money matters. He sums up the significance of all the great events of the year, bearing upon human progress in general, and upon each specific enterprise in particular. With palatial mansions, parks, and farms great and small, scattered through several counties, he is the greatest radical in England. He distances the Chartists altogether in his programme, and adds several new points to their political creed. He not only advocates manhood suffrage, but womanhood suffrage, and woman-seats in Parliament. Then he is a great friend of a reform which the Chartists grievously overlook, and which would make thousands of them voters if they would adopt it. That is, Total Abstinence from Tobacco, as well as from Ardent Spirits. Thus, no report of modern times equals the good Squire's summing-up, which he gives on these occasions, from the great farm-wagon tribune, to the multitudinous and motley congregation assembled under his park trees. This year it was unusually rich and piquant, from the expanded area of events and aspects. In presenting these, as bearing upon the causes of Temperance, Peace, Anti-War, Anti-Slavery, Anti-Tobacco, Anti-Capital Punishment, Anti-Church-Rates, Free Trade, Woman's Rights, Parliamentary Reform, Social Reform, Scientific Progress, Discovery of the Sources of the Nile, and other important movements, he was necessarily obliged to be somewhat discursive. But he generalised with much ease and perspicuity, and conducted the thread of his discourse, like a rivulet of light, through the histories of the year; transporting the mind of his audience from doings in Japan to those in America, from Poland to Mexico, and through stirring regions of Geography, Politics, Philanthropy, Social Science and Economy, by gentle and interesting transitions. This annual statement is very valuable and instructive, and should have a wider publicity than it usually obtains.

When "the fine old English gentleman all of the olden time" has concluded his *resume* of the year's progress, and the prospects it leaves to the one incoming, the orators of the different causes which he has thus reported, arise one after the other, and the bright air and the green foliage of the over-spreading trees, as well as the listening multitude below are stirred with fervid speeches, sometimes interspersed with "music from the band." The Festival is wound up by a banquet in the hall, given by the munificent host to a large number of guests, representing the various good movements advocated from the platform described. Many Americans have spoken from that rostrum, and sat at that banquet table in years gone by, and they will attest to the correctness of these slight delineations of the character of the host and of the annual festival that will perpetuate his name in long and pleasant remembrance.

Oakham is a goodly and pleasant town, the chief and capital of Rutland-shire. It has the ruins of an old castle in its midst, and several interesting antiquities and customs. It, too, has its unique speciality or prerogative. I was told that every person of title driving through the town, or coming to reside within the jurisdiction of its bye-laws, must leave his card to the authorities in the shape of a veritable *horse-shoe*. It is said that the walls of the old town hall are hung with these iron souvenirs of distinguished visits; thus constituting a museum that would be instructive to a farrier or blacksmith, as well as to the antiquarian.

From Oakham I walked to Melton Mowbray, a cleanly, good-looking town in Leicestershire, situated on the little river Eye. One cannot say exactly in regard to Rutlandshire what an Englishman once said to the authorities of a pigmy Italian duchy, who ordered him to leave it in twenty-four hours. "I only require fifteen minutes," said cousin John, with a look and tone which Jonathan could not imitate. This rural county is to the shire-family of England what Rhode Island is to the American family of States—the smallest, but not least, in several happy characteristics.

I spent a quiet Sabbath in Melton Mowbray; attended divine service in the old parish church and listened to two extemporaneous sermons full of simple and earnest teaching, and delivered in a conversational tone of voice. Here, too, the parish church was seated in the midst of the great congregation which had long ceased to listen to the call of its Sabbath bells. It was a beautiful and touching arrangement of the olden time to erect the House of Prayer in the centre of "God's Acre," that the shadow of its belfry and the Sabbath voice of its silvery bells might float for centuries over the family circles lying side by side in their long homes around the sanctuary. There was a good and tender thought in making up this sabbath society of the living and the dead; in planting the narrow pathway between the two Sions with the white milestones of generations that had travelled it in ages gone, leaving here and there words of faith, hope and admonition to those following in their footsteps. It is one of the contingencies of "higher civilization" that this social economy of the churchyard, that linked present and past generations in such touching and instructive companionship, has been suspended and annulled.

Melton Mowbray has also a very respectable individuality. It is a great centre for the scarlet-coated Nimrods who scale hedges and ditches, in well-mounted squadrons, after a fox *preserved* at great expense and care to become the victim of their valor. But this is a small and frivolous distinction compared with its celebrated manufacture of *pork-pies*. It bids fair to become as famous for them as Banbury is for buns. I visited the principal establishment for providing the travelling and picnicking world with these very substantial and palatable portables. I went under the impulse of that uneasy, suspicious curiosity to peer into the forbidden mysteries of the kitchen which generally brings no satisfaction when gratified, and which often admonishes a man not only to eat what is set before him without any questions for conscience sake, but also for the sake of the more delicate and exacting sensibilities of the stomach. I must confess my first visit to this, the greatest pork-pie factory in the world, savored a little of the anxiety to know the worst, instead of the best, in regard to the solid materials and lighter ingredients which entered into the composition of these suspiciously cheap luxuries. There were points also connected with the process of their elaboration which had given me an undefinable uneasiness in the refreshment rooms of a hundred railway stations. I was determined to settle these moot points once for all. So I entered the establishment with an eye of as keen a speculation as an exciseman's searching a building for illicit distillery, and I came out of it a more charitable and contented man. All was above board, fair and clean. The meat was fresh and good. The flour was fine and sweet; the butter and lard would grace the neatest housewife's larder; the forms on which the pies were moulded were as pure as spotless marble. The men and boys looked healthy and bright; their hands were smooth and clean, and their aprons white as snow. Not one of them smoked or took snuff at his work. I saw every process and implement employed in the construction of these pies for the market; the great tubs of pepper and spice, the huge ovens, the cooling racks, the packing room; in a word, every department and feature of the establishment. And the best thing that I can say of it is this: that I shall eat with better satisfaction and relish hereafter the pies bearing the brand of Evans, of Melton Mowbray, than I ever did before. The famous Stilton cheese is another speciality of this quiet and interesting town, or of its immediate neighborhood. So, putting the two articles of luxury and consumption together, it is rather ahead of Banbury with its cakes.

On Monday, August 11th, I resumed my walk northward, and passed through a very highly cultivated and interesting section. About the middle of the afternoon, I reached Broughton Hill, and looked off upon the most beautiful and magnificent landscape I have yet seen in England. It was the Belvoir Vale; and it would be worth a hundred miles' walk to see it, if that was the only way to reach it. It lay in a half-moon shape, the

base line measuring apparently about twenty miles in length. As I sat upon the high wall of this valley, that overlooks it on the south, I felt that I was looking upon the most highly-finished piece of pre-Raphaelite artistry that could be found in the world,—the artistry of the plough, glorious and beautiful with the unconscious and involuntary pictures which patient human labor paints upon the canvas of Nature. Never did I see the like before. If Turner had the shaping of the ground entirely for an artistic purpose, it could not have been more happily formed for a display of agricultural pictures. What might be called the *physical* vista made the most perfect *hemiorama* I ever looked upon. The long, high, wooded ridge, including Broughton Hill, *eclipsed*, as it were, just half the disk of a circle twenty miles in diameter, leaving the other half in all the glow and glory that Nature and that great blind painter, Agricultural Industry, could give to it. The valley with its foot against this mountainous ridge, put out its right arm and enfolded to its bosom a little, beautiful world of its own of about fifty miles girth. In this embrace were included hundreds of softly-rounded hills, with their intervening valleys, villages, hamlets, church spires and towers, plantations, groves, copses and hedge-row trees, grouped by sheer accident as picturesquely as Turner himself could have arranged them. The elevation of the ridge on which I sat softened down all these distant hills, so that they looked only like little undulating risings by which the valley gently ascended to the blue rim of the horizon on the north.

It was an excellent standpoint on which to balance Nature and Human Industry; to estimate their separate and joint work upon that vast landscape. A few centuries ago, perhaps about the time that the Mayflower sighted Plymouth Rock, this valley, now so indescribably beautiful, was almost in the state of nature. Wolves and wild boars may have been prowling about in the woods and tangled thickets that covered this ridge back for several leagues. Bushes, bogs and briers, and coarse

prairie grass roughened the bottom of this valley; matted heather, furze, broom and clumps of shrubby trees, all those hills and uplands arising in the background to the northward horizon. This declining sun, and the moon and stars that will soon follow in the pathway of its chariot, like a liveried *cortege*, shone upon that scene with all the light they will give this day and night. The rain and dew, and all the genial ministries of the seasons, did their unaided best to make it lovely and beautiful. The sweetest singing-birds of England came and tried to cheer its solitude with their happy voices. The summer breezes came with their softest breath, whispering through brake, bush and brier the little speeches of Nature's life. The summer bees came and filled all those heather-purpled acres with their industrial lays, and sang a merry song in the door of every wild-flower that gave them the petalled honey of its heart. All the trained and travelling industrials and all the sweet influences of Nature came and did all they could without man's help to make this great valley most delightful to the eye. But the wolves still prowled and howled; the briers grew rough and rank; the grass, coarse and thin; the heathered hills were oozy and cold in their watery beds; the clumpy, shrubby trees wore the same ragged coats of moss; and no feature of the scene mended for the better from year to year.

Then came the great Blind Painter, with his rude, iron pencils, to the help of Nature. He came with the Axe, Plough and Spade, her mightiest allies. With these he had driven wild Druidic Paganism back mile by mile from England's centre; back into her dark fastnesses. With the Axe, Spade and Plough he chased the foul beasts and barbarisms from the island. Two centuries long was he in painting this Beautiful Valley. Nature ground and mixed the colors for him all the while, for he was blind. He was poor; often cold and hungry, and his children, with blue fingers and pale, silent eyes, sometimes asked for bread in winter he could not give. He lived in a low cottage, small, damp and dark, and laid him down at night upon a bed of straw. He could not read; and his thoughts of human life and its hereafter were few and small. He had no taste for music, and seldom whistled at his work. He wore a coarse garment, of ghostly pattern, called a smock-frock. His hat just rounded his head to a more globular and mindless form. His shoes were as heavy as a horse's with iron nails. He had no eye nor taste for colors. If all the trees, if all the crops of grain, grass and roots on which he wrought his life long, had come out in brickdust and oil, it would have been all the same to him, if they had sold as high in the market, and beer and bread had been as cheap for the uniformity. And yet he was the Turner of this great painting. He is the artist that has made England a gallery of the finest agricultural pictures in the world. And in no country in Christendom is High Art so appreciated to such pecuniary patronage and valuation as here. In none is the genius of the Pencil so treasured, so paid, and almost worshipped as here. The public and private galleries of Britain hold pictures that would buy every acre of the island at the price current of it when Elizabeth was queen. One of Turner's landscapes would pay for a whole Highland county at its valuation when Mary held her first court at Holyrood.

I sit here and look off upon this largest, loveliest picture the Blind Painter has given to England. I note his grouping of the ivy-framed fields, of every size and form, panelling the gently-rounded hills, and all the soft slopes down to the foot of the valley; the silvery, ripe barley against the dark-green beans; the rich gold of the wheat against the smooth, blue-dashed leaves of the mangel wurzel or rutabaga; the ripening oats overlooking a foreground of vividly green turnips, with alternations of pasture and meadow land, hedges running in every direction, plantations, groves, copses sprinkled over the whole vista, as if the whole little world, clear up to the soft, blue fringe of the horizon, were the design and work of a single artist. And this, and ten thousand pictures of the same genius, were the work

of the Briarean-handed BLIND PAINTER, who still wears a smock-frock and hob-nailed shoes, and lives in a low, damp cottage, and dines on bread and cheese among the golden sheaves of harvest!

O, Mother England! thou that knightest the artists while living, and buildest their sepulchres when dead; thou that honorest to such stature of praise the plagiarists upon Nature, and clothest the copyists of patient Labor's pictures in such purple and fine linen; thou whose heart is softening to the sweet benevolences of Christian charity in so many directions,—wilt thou not think, with a new sentiment of kindness and sympathy, on this Blind Painter, who has tapestried the hills and valleys of thy island with an artistry that angels might look upon with admiration and wonder!

Wilt thou not build him a better cottage to live in?

Wilt thou not give him something better than dry bread and cold bacon for dinner in harvest?

Wilt thou not teach all his children to read the alphabet and the blessed syllables of the Great Revelation of God's Love to man?

Wilt thou not make a morning-ward door in his dwelling and show him a future with a sun in it, in *this* world, as well as the world to come?

Wilt thou not open up a pathway through the valley of his humiliation by which his children may ascend to the better conditions of society?

CHAPTER XIV.

NOTTINGHAM AND ITS CHARACTERISTICS—NEWSTEAD ABBEY—MANSFIELD—TALK IN A BLACK-SMITH'S SHOP—CHESTERFIELD, CHATSWORTH AND HADDON HALL—ARISTOCRATIC CIVILISATION, PRESENT AND PAST.

From the Belvoir Vale I continued my walk to Nottingham the following day; crossing a grand old bridge over the Trent. Take it all in all, this may

be called perhaps the most English town in England; stirring, plucky and radical; full of industrial intellect and vigor. Its chief businesses involve and exercise thought; and thought educed into one direction and activity, runs naturally into others. The whole population, under these influences, has become *peopled* to a remarkable status and strength of opinion, sentiment and action. They prefix that large and generous quality to their best doings and institutions, and have their Peoples' College, Peoples' Park, etc. The Peoples' Charter had its stronghold here, and all radical reforms are sure to find sympathy and support among the People of Nottingham. I should think no equal population in the kingdom would sing "Britons never, never will be slaves," with more spirit, or, perhaps, with more understanding. Their plucky, English natures became terribly stirred up in the exciting time of the Reform Bill, and they burned down the magnificent palace-castle of the old Duke of Newcastle, crowning the mountainous rock which terminates on the west the elevated ridge on which the town is built. When the Bill was carried, and the People had cooled down to their normal condition of mind, they were obliged to pay for this evening's illumination of their wrath pretty dearly. The Duke mulcted the town and county to the tune of £21,000, or full $100,000. The castle was no Chepstow structure, rough and rude for war, but more like the ornate and castellated palace at Heidelberg, and it was almost as high above the Trent as the latter is above the Neckar. The view the site commands is truly magnificent, embracing the Trent Valley, and an extensive vista beyond it. It was really the great lion of the town, and the People, having paid the £21,000 for dismounting it, because it roared in the wrong direction on the Reform Bill, expected, of course, that His Grace the Duke would set it up again on the old pedestal, with its mane and tail and general aspect much improved. But they counted without their host. "Is it not lawful to do what I will with my own," was the substance of his reply; and there stands the blackened, crumbling ruin to this day, as a silent but grim reproach to the People for letting their angry passions rise to such destructive excitement on political questions.

Hosiery and lace are the two great manufacturing interests of Nottingham, and the tons of these articles it turns out yearly for the world are astonishing in number and value. A single London house employs 3,000 hands in the town and immediate vicinity upon hosiery alone for its establishment. Lace now seems to lead the way, and there are whole streets of factories and warehouses busy with its manufacture and sale. Perhaps no fabric in the world ever tested the ingenuity and value of machinery like this. The cost has been reduced, from the old hand-working to the present process, from three dollars to three cents a yard! I think no machinery yet invented has been endowed with more delicate functions of human reason and genius than that employed upon the flower-work of this subtle drapery. Until I saw it with my own eyes, I had concluded that the machinery invented or employed in America for setting card-teeth was the most astute, and as nearly approaching the faculties of the human mind in its apparent thought-power, as it was reverent and safe to carry anything made of iron and steel, or made by man at all. To construct a machine which should pass between its fingers a broad belt of leather and a fine thread of wire, prick rows of holes across the breadth of the leather, bend, cut off, and insert the shank ends of the teeth clear through these holes, and clinch them on the back side, and pour out a continuous, uninterrupted stream of perfectly-teethed belt, all ready for carding,— this, I fancied, was the *ne plus ultra* of mechanical inventions. But it is quite surpassed by the lace-weaving looms of Nottingham, that work out, to exquisite perfection, all the flowers, leaves, vines and vein-work of nature. It was wonderful to see the ductility of cotton, as here exemplified. The *bobbins*, which, I suppose, are a mere refinement upon the old hand-thrown shuttle, are of brass, about the size of half-a-crown. A

groove that will just admit the thin edge of a case-knife, is cut into the rim of the little wheel, about one quarter of an inch deep. A cotton thread, 120 yards in length, and strong enough to be twitched about and twisted by a score of vigorous, chattering, iron fingers, is wound around in this groove. But it would be idle to attempt a description of either the machinery or the process.

I went next into a large establishment for dyeing, dressing, winding and packing the lace for market. It was startling to see the acres of it dyed black for mourning. Really there seemed enough of it to drape the whole valley of the shadow of death! It was an impressive sight truly. If there were other establishments doing the same thing, Nottingham must turn out weeds of grief enough for several millions of mourning widows, mothers, sisters and daughters in a year. I ascended into the dressing-room, I think they called it, in the upper story, where there was a piece containing one twenty-fifth of an acre of lace undergoing a fearful operation for a human constitution to sustain. It was necessary that the heat of the apartment should be kept at *one hundred and twenty* degrees! There was a large number of women and girls, and a few men and boys working under this melting ordeal. And one of the proprietors was at their head, in a rather summer dress, and with a seethed and crimson face beaded with hot perspiration. It was a very delicate and important operation which he had not only to watch with his own eyes, but to work at with his own hands. I was glad to learn that he was a staunch Protestant, and did not believe in *purgatory*; but those poor girls!—could they be expected to hold to the same belief under such a test?

I was told that they could get up lace so cheap that the people of the town frequently cover their gooseberry bushes with it to keep off the insects. Spider-webbing is a scarcely more gossamer-like fabric. Sixteen square yards of this lace only weigh about an ounce! If the negroes on one of the South Carolina Sea-island plantations could have been shut into that dressing-room for two

whole minutes, with the mercury at 120 degrees, they would have rolled up the whites of their eyes in perfect amazement and made a rush for "Dixie" again.

From Nottingham I made an afternoon walk to Mansfield. The weather was splendid and the country in all the glory of harvest. On reaching Newstead Abbey, I found, to my regret, that the *entree* to the public had been closed by the new proprietor, one, I was told, of the manufacturing gentry of the Manchester school. Not that he was less liberal and accommodating to sight-seers than his predecessors, but because he was making very extensive and costly improvements in the buildings and grounds. I have seen nothing yet in England to compare, for ornate carving, with the new gate-way he is making to the park. It is of the finest kind of arabesque work done in stone that much resembles the Caen. This prevention barred me from even a distant view of the once famous residence of Lord Byron, as it could not be seen from the public road.

Within about three miles of Mansfield, I came to a turnpike gate,—a neat, cozy, comfortable cottage, got up in the Gothic order. I stopped to rest a moment, and noticing the good woman setting her tea-table, I invited myself to a seat at it, on the inn basis, and had a pleasant meal and chat with her and an under-gamekeeper of the Duke of Portland, who had come in a little before me. The stories he told me about the extent of the Duke's possessions were marvellous, more especially in reference to his game preserves. I should think there must be a larger number of hares, rabbits and partridges on his estate than in the whole of New England. As I sat engaged in conversation with the woman of the house and this accidental guest, an unmistakable American face met my eyes, as I raised them to the opposite wall. It was the familiar face of a Bristol clock, made in the Connecticut village adjoining the one in which I was born. It wore the same honest expression, which a great many ill-natured people, especially in our Southern States, have regarded as covering a dishonest and untruthful mind, or a bad memory of the hours. Still it is the most ubiquitous *Americanism* in the world, and it is pleasant to see its face in so many cottages of laboring men from Land's End to John O'Groat's.

Mansfield is a very substantial and venerable town, bearing a name which one distinguished man has rendered illustrious by wearing it through a brilliant life. It is situated near the celebrated Sherwood Forest, and is marked by many features of peculiar interest. One of its noticeable celebrities is the house in which Lord Chesterfield resided. It is now occupied by a Wesleyan minister, who elaborates his sermons in the very room, I believe, in which that fashionable nobleman penned his polite literature for youthful candidates for the uppermost circles of society. In the centre of the market place there is a magnificent monument erected to the memory of the late Lord George Bentinck, who was held in high esteem by the people of the town and vicinity. The manufactures are pretty much the same as in Nottingham. They turn out a great production of raw material in red sandstone, very much resembling our Portland, quite as fine, hard and durable. Immense blocks of it are quarried and conveyed to London and to all parts of the kingdom. The town also supplies a vast amount of moulding sand, of nearly the same color and consistency as that we procure from Albany. I stopped on my way into the town to take a turn through the cemetery, which was very beautifully laid out, and looked like a great garden lawn belted with shrubbery, and illuminated with the variegated lamps of flowers of every hue and breath. The meandering walks were all laid with asphalte, which presented a new and striking contrast to the gorgeous borders and the vivid green of the cleanly shaven grass. Many of the little graves were made in nests of geraniums and other modest and sweet-eyed stars of hope.

Next day I had a very enjoyable walk in a north-westerly direction to Chesterfield. On the way, called in at a blacksmith's shop, and had a long talk with the smith-in-chief on matters connected with his trade. The "custom-work" of such shops in country villages in England is like that in ours fifty years ago—embracing the greatest variety of jobs. Articles now made with us in large manufacturing establishments at a price which would starve a master and his apprentice to compete with, are hammered out in these English shops on a single anvil. On comparing notes with this knight of the hammer, I learned a fact I had not known before. His price for horse-shoeing varied according to the size of the hoof, just as our leather-shoemakers charge according to the foot. On taking leave of him he intimated, in the most frank and natural way in the world, that, in our exchange of information, the balance was in his favor, and that I could not but think it fair to pay him the difference. I looked at him first inquiringly and doubtingly, embarrassed with the idea that I had not understood him, or that he was a journeyman and not the master of the establishment. But he was as free and easy and natural as possible. An American tobacco-chewer, of fifty years' standing, would not have asked a cut from a neighbor's "lady's twist," or "pig-tail" in more perfect good faith. That good, round, English face would have blushed crimson if the man suspected that I misunderstood him. Nay, more, he would quite likely have thrown the pennies at my head if I had offered them to him to buy bread or bacon with for himself and family. I had no reason for a moment's doubt. It all meant *beer*, "only that and nothing more;" a mere *pour boire* souvenir to celebrate our mutual acquaintance. So I gave him a couple of pennies, just as I would have given him a bite of tobacco if we had both been in that line. I feared to give him more, lest he might think I meant bread and bacon and thought him a beggar. But I ventured to tell him, however, that I did not use that beverage myself, and hoped he would wish me health in some better enjoyment.

I saw, for the first time, a number of Spanish cattle feeding in a pasture. They were large, variously colored animals with the widely-branching horns that distinguish them. A man must have a long range of buildings to stable a score of creatures with such horns, and for that reason they will only be kept as curiosities in these northern latitudes. And they are curiosities of animal life, heightened to a wonderment when placed side by side with the black Galloways, or those British breeds of cattle which have no horns at all. I should not wonder, however, if this large, cream-colored stock from Spain should be introduced here to cross with the Durhams, Devons, and Herefords.

When about half-way from Mansfield to Chesterfield, a remarkable change came over the face of the landscape. The mosaic work of the hill-sides and valleys showed more green squares than before. Three-fourths of the fields were meadow or pasture, or in mangel or turnips. There was but one here and there in wheat or other grain. The road beneath and the sky above began to blacken, and the chimneys of coal-pits to thicken. Sooty-faced men, horses and donkeys passed with loaded carts; and all the premonitory aspects of the "black country" multiplied as I proceeded. I do not recollect ever seeing a landscape change so suddenly in England.

Chesterfield is an intelligent looking town, evidently growing in population and prosperity. It has its own unique speciality; almost as strikingly distinctive as that of Strasburg or Pisa. This is the most ambiguous and mysterious church spire in the world. It would be very difficult to convey any idea of it by any description from an unaided pen; and there is nothing extant that would avail as an illustration. The church is very old and large, and stands upon a commanding eminence. The massive tower supports a tall but suddenly tapering spire of the most puzzling construction to the eye. It must have been designed by a monk of the olden time, with a Chinese turn of ingenuity. There is no order known to architecture to furnish a term or likeness for it. A ridgy, spiral spire are the three most descriptive words, but these are not half enough for stating the shape, style and posture of this strange steeple. It is difficult even to assist the imagination to form an idea of it. I will essay a few words in that direction. Suppose, then, a plain spire, 100 feet high, in the form of an attenuated cone, planted upon a heavy church tower. Now, in imagination, plough this cone all around into deep ridges from top to bottom. Then mount to the top, and, with a great iron wrench, give it an even twist clear down to the base, so that each ridge shall wind entirely around the spire between the bottom and the top. Then, in giving it this screw-looking twist, bend over the top, with a gentle incline all the way down, so that it shall be "out of perpendicular" by about three feet. Then come down and look at your work, and you will be astonished at it, standing far or near. The tall, ridgy, curved, conical screw puzzles you with all sorts of optical illusions. As the eyes in a front-face portrait follow you around the room in which it is hung, so this strange spire seems to lean over upon you at every point, as you walk round the church. Indeed, I believe it was only found out several centuries after its erection, that it absolutely leaned more in one direction than another. It is a remarkable sight from the railway as you approach the town from a distance. If it may be said reverently, the church, standing on comparatively a hill, not only lifts its horn on high, but one like that of a rhinoceros, considerably curved. Just outside the town stands the house in which George Stephenson lived his last days, and ended his great life of benefaction to mankind; leaving upon that haloed spot a *biograph* which the ages of time to come shall not wash out.

From Chesterfield I diverged westward to see Chatsworth and Haddon Hall. Whoever makes this walk or ride, let him be sure to stop at Watch Hill on the way, and look at the view eastward. It is grander than that of Belvoir Vale, if not so beautiful.

It was a pleasure quite equal to my anticipation to visit Chatsworth for the first time, after a sojourn in England, off and on, for sixteen years. It is the lion number three, according to the American ranking of the historical edifices and localities of England. Stratford-upon-Avon, Westminster Abbey and Chatsworth are the three representative celebrities which our travellers think they must visit, if they would see the life of England's ages from the best standpoints. And this is the order in which they rank them. Chatsworth and Haddon Hall should be seen the same day if possible; so that you may carry the impressions of the one fresh and active into the other. They are the two most representative buildings in the kingdom. Haddon is old English feudalism *edificed*. It represents the rough grandeur, hospitality, wassail and rude romance of the English nobility five hundred years ago. It was all in its glory about the time when Thomas-a-Becket the Magnificent used to entertain great companies of belted knights of the realm in a manner that exceeded regal munificence in those days,—even directing fresh straw to be laid for them on his ample mansion floor, that they might not soil the bravery of their dresses when they bunked down for the night. The building is brimful of the character and history of that period. Indeed, there are no two milestones of English history so near together, and yet measuring such a space of the nation's life and manners between them, as this hall and that of Chatsworth. It was built, of course, in the bow-and-arrow times, when the sun had to use the same missiles in shooting its barbed rays into the narrow apertures of old castles—or the stone coffins of fear-hunted knights and ladies, as they might be called. What a monument this to the dispositions and habits of the world, outside and inside, of that early time! Here is the porter's or warder's lodge just inside the huge gate. To think of a living being with a human soul in him burrowing in such a place!—a big, black sarcophagus without a lid to it, set deep in the solid wall. Then there is the chapel. Compare it with that of Chatsworth, and you may count almost on your fingers the centuries that have intervened between them. It was new-roofed soon after the discovery of America, and perhaps done up to some show of decency and comfort. But how small and rude the pulpit and pews—looking like rough-boarded potato-bins! Here is the great banquet-hall, full to overflowing with the tracks and cross-tracks of that wild, strange life of old. There is a fire-place for you, and a mark in the chimney-back of five hundred Christmas logs. Doubtless this great stone pavement of a floor was carpeted with straw at these banquets, after the illustrious Becket's pattern. Here is a memento of the feast hanging up at the top of the kitchenward door;—a pair of roughly-forged, rusty handcuffs amalgamated into one pair of jaws, like a musk-rat trap. What was the use of that thing, conductor? "That, sir, they put the 'ands in of them as shirked and didn't drink up all the wine as was poured into their cups, and there they made them stand on tiptoe up against that door, sir, before all the company, sir, until they was ashamed of theirselves." Descend into the kitchen, all scarred with the tremendous cookery of ages. Here they roasted bullocks whole, and just back in that dark vault with a slit or two in it for the light, they killed and dressed them. There are the relics of the shambles. And here is the great form on which they cut them up into manageable pieces. It would do you good, you Young America, to see that form, and the cross-gashes of the meat-axe in it. It is the half of a gigantic English oak, which was growing in Julius Caesar's time, sawed through lengthwise, making a top surface several feet wide, black and smooth as ebony. Some of the bark still clings to the under side. The dancing hall is the great room of the building. All that the taste, art and wealth of that day could do, was done to make it a splendid apartment, and it would pass muster still as a comfortable and respectable *salon*. As we pass out, you may decipher the short prayer cut in the wasting stone of a side portal, "GOD SAVE THE VERNONS!" I hope this prayer has been favorably answered; for history records much virtue in the family, mingled with some romantic *escapades*, which have contributed, I believe, to the entertainment of many novel readers.

Just what Haddon Hall was to the baronial life and society of England five hundred years ago, is Chatsworth to the full stature of modern civilization and aristocratic wealth, taste and position. Of this it is probably the best measure and representative in the kingdom; and as such it possesses a special value and interest to the world at large. Were it not for here and there such an establishment, we should lack waymarks in the progress of the arts, sciences and tastes of advancing civilization. Governments and joint-stock companies may erect and fill, with a world of utilities and curiosities of ancient and modern times, British Museums, National Galleries, Crystal Palaces and Polytechnic Institutions; but not one of these, nor the Louvre, nor Versailles, nor the Tuileries can compete with one private mind, taste and will concentrated upon one great work for a lifetime, when endowed with the requisite perceptions and means competent to carry that work to the highest perfection of science, genius and art. Museums, galleries and public institutions of art are exclusively *visiting* places. The elegancies of *home* life are all shut out of their attractions. You see in them the work and presence of a committee, or corporation, often in discrepant layers of taste and plan. One mind does not stand out or above the whole, fashioning the *tout-ensemble* to the symmetrical lines of one governing, all-pervading and shaping thought. You see no exquisite artistry of drawing-room or boudoir elegance and luxury running through living apartments of home, out into the conservatories, lawns, gardens, park and all its surroundings and embellishments, making the whole like a great illuminated volume of family life, which you may peruse page by page, and trace the same pen and the same story from beginning to end. Even the grandest royal residences lack, in this quality, what you will find at Chatsworth. They all show the sharp-edged strata of unaffiliated tastes and styles of different ages and artists. They lack the oneness of a single individuality, of one great symmetrical conception.

This one-mindedness, this one-man power of conception and execution gives to the Duke of Devonshire's palace at Chatsworth an interest and a value that probably do not attach to any other private establishment in England. In this felicitous characteristic it stands out in remarkable prominence and in striking contrast with nearly all the other baronial halls of the country. It is the parlor pier-glass of the present century. It reflects the two images in vivid apposition—the brilliant civilization of this last, unfinished age in which we live and the life of bygone centuries; that is, if Haddon Hall shows its face in it, or if you have the features of that antiquity before your eyes when you look into the Chatsworth mirror. The whole of this magnificent establishment bears the impress of the nineteenth century, inside and outside. The architecture, sculpture, carving, paintings, engravings, furniture, libraries, conservatories, flowers, shrubberies and rockeries all bear and honor the finger-prints of modern taste and art. In no casket in England, probably, have so many jewels of this century's civilization been treasured for posterity as in this mansion on the little meandering Derwent. If England has no grand National Gallery like the French Louvre, she has works of art that would fill fifty Louvres, collected and treasured in these quiet private halls, embosomed in green parks and plantations, from one end of the land to the other. And in no other country are the private treasure-houses of genius so accessible to the public as in this. They doubtless act as educational centres for refining the habits of the nation; exerting an influence that reaches and elevates the homes of the people, cultivating in them new perceptions of beauty and comfort; diffusing a taste for embowering even humble cottages in shrubbery; making little flower-fringed lawns, six feet by eight or less; rockeries and ferneries, and artificial ruins of castles or abbeys of smaller dimensions still.

In passing through the galleries and gardens of Chatsworth you will recognise the originals of many works of art which command the admiration of the world. The most familiar to the American visitor will probably be the great painting of the Bolton Abbey Scene, the engravings of which are so numerous and admired on both sides of the Atlantic. But there is the original of a greater work, which has made the wonder of the age. It is the original of the Great Crystal Palace of 1851, and the mother of all the palaces of the same structure which have been or will be erected in time past or to come. Here it diadems at Chatsworth the choice plants and flowers of all the tropics; presenting a model which needed only expansion, and some modifications, to furnish the reproduction that delighted the world in Hyde Park in 1851.

I was pleasantly impressed with one feature of the economy that ruled at Chatsworth. Although there were between one and two thousand deer flecking the park, it was utilised to the pasture of humbler and more useful animals. Over one hundred poor people's cows were feeding demurely over its vast extent, even to the gilded gates of the palace. They are charged only £2 for the season; which is very moderate, even cheaper than the stony pasturage around the villages of New England. I noticed a flock of Spanish sheep, black-and-white, looking like a drove of Berkshire hogs, and seemingly clothed with bristles instead of wool. They are kept rather as curiosities than for use.

Chatsworth, with all its treasures and embodiments of wealth, art and genius, with an estate continuous in one direction for about thirty miles, is but one of the establishments of the Duke of Devonshire. He owns a palace on the Thames that might crown the ambition of a German prince. He also counts in his possessions old abbeys, baronial halls, parks and towns that once were walled, and still have streets called after their gates. If any country is to have a personage occupying such a position, it is well to have a considerable number of the same class, to *yeomanise* such an aristocracy—to make each feel that he has his peers in fifty others. Otherwise an isolated duke would have to live and move outside the pale of human society; a proud, haughty entity dashing about, with not even a comet's orbit nor any fixed place in the constellation of a nation's communities. It is of great necessity to him, independent of political considerations, that there is a House of Peers instituted, in which he may find his social level; where he may meet his equals in considerable numbers, and feel himself but a man.

CHAPTER XV.

SHEFFIELD AND ITS INDIVIDUAL-ITY—THE COUNTRY, ABOVE GROUND AND UNDER GROUND—WAKEFIELD AND LEEDS—WHARF VALE—FARNLEY HALL—HARRO-GATE; RIPLEY CASTLE; RIPON; CONSERVATISM OF COUNTRY TOWNS—FOUNTAIN ABBEY; STUD-LEY PARK—RIEVAULX ABBEY—LORD FAVERSHAM'S SHORT-HORN STOCK.

From Chatsworth I went on to Sheffield, crossing a hilly moorland belonging to the Duke of Rutland, and containing 10,000 acres in one solid block. It was all covered with heather, and kept in this wild, bleak condition for game. Here and there well-cultivated farms, as it were, bit into this cold waste, rescuing large, square morsels of land, and making them glow with the warm flush and glory of luxuriant harvests; thus showing how such great reaches of desert may be made to blossom like the rose under the hand of human labor.

Here is Sheffield, down here, sweltering, smoking, and sweating, with face like the tan, under the walls of these surrounding hills. Here live and labor Briareus and Cyclops of modern mythology. Here they—

Swing their heavy sledge,
With measured beats and slow;
Like the sexton ringing the village bell,
When the evening sun is low.

Here live the lineal descendants of Thor, christianised to human industries.

Here the great hammer of the Scandinavian Thunderer descended, took nest, and hatched a brood of ten thousand little iron beetles for beating iron and steel into shapes and uses that Tubal Cain never dreamed of. Here you may hear their clatter night and day upon a thousand anvils. O, Vale of Vulcan! O, Valley of Knives! Was ever a boy put into trousers, in either hemisphere, that did not carry in the first pocket made for him one of thy cheap blades? Did ever a reaper in the Old World or New cut and bind a sheaf of grain, who did not wield one of thy famous sickles? All Americans who were boys forty years ago, will remember three English centres of peculiar interest to them. These were Sheffield, Colebrook Dale, and Paternoster Row. There was hardly a house or log cabin between the Penobscot and the Mississippi which could not show the imprint of these three places, on the iron tea-kettle, the youngest boy's Barlow knife, and his younger sister's picture-book. To the juvenile imagination of those times, Sheffield was a huge jack-knife, Colebrook Dale a porridge-pot, and Paternoster Row a psalm-book, each in the generative case. How we young reapers used to discuss the comparative merits and meanings of those mysterious letters on our sickles, B.Y and I.R! What were they? Were they beginnings of words, or whole words themselves? Did they stand for things, qualities, or persons? "Mine is a *By* sickle; mine is an *Ir* one. Mine is the best," says the last, "for it has the finest teeth and the best curve." That was our boys' talk in walking through the rye, with bent backs and red faces, a little behind our fathers; who cut a wider work to enable us to keep near them.

In what blacksmith shop or hardware house in America does not Sheffield show its face and faculties? Did any American, knowing the difference between cast-iron and cast-steel, ever miss the sight of Naylor and Sanderson's yellow labels in his travels? How many millions of acres of primeval forest have the ages edged with their fine steel cut through, and given to the plough! Fashion has its Iron Age as well as its Gold-

en; and, what is more remarkable, the first of the two has come last, in the fitful histories of custom. And this last freak of feminine taste has brought a wonderful grist of additional business to the Sheffield mill. The fair Eugenie has done a good thing for this smoky town, well deserving of a monument of burnished steel erected to her memory on one of these hills. More than this; as Empress of Crinoline, she should wear the iron crown of Charlemagne in her own right. Her husband's empire is but a mere *arrondissement* compared with the domain that does homage to her sceptre. Sheffield is the great arsenal of her armaments. Sheffield cases ships of war with iron plates a foot thick; but that is nothing, in pounds avoirdupois, compared with the weight of steel it spins into elastic springs for casing the skirts of two hundred millions of the fair Eugenie's sex and lieges in the two hemispheres. It is estimated that ten thousand tons of steel are annually absorbed into this use in Christendom; and Sheffield, doubtless, furnishes a large proportion of it.

Here I had another involuntary walk, not put down in the programme of my expectations. On inquiring the way to Fir Vale, a picturesque suburb where a friend resided, I was directed to a locality which, it was suggested, must be the one I meant, though it was called Fir View. I followed the direction given for a considerable distance, when it was varied successively by persons of whom I occasionally inquired. After ascending and descending a number of steep hills, I suddenly came down upon the town again from the south, having made a complete circuit of it; a performance that cost me about two hours of time and much unsatisfactory perspiration. Fearing that a second attempt would be equally unsuccessful, I took the Leeds road, and left the Jericho at the first round. Walked about nine miles to a furnace-lighted village called very appropriately Hoyland, or Highland, when anglicised from the Danish. It commands truly a grand view of wooded hills and deep valleys dashed with the sheen of ripened grain.

The next day I passed through a good sample section of England's wealth and industry. Mansions and parks of the gentry, hill, valley, wheat-fields, meadows of the most vivid green; crops luxuriant in most picturesque alternations; in a word, the whole a vista of the richest agricultural scenery. And yet out of the brightest and broadest fields of wheat, barley and oats, towered up the colliery chimneys in every direction, like good-natured and swarthy giants smoking their pipes complacently and "with comfortable breasts" in view of the goodly scene. The golden grain grew thick and tall up to the very pit's mouth. In the sun-light above and gas-light below human industry was plying its differently-bitted implements. There were men reaping and studding the pathway of their sickles through the field with thickly-planted sheaves. But right under them, a hundred fathoms deep, subterranean farmers were at work, with black and sweaty brows, garnering the coal-harvest sown there before the Flood. Sickle above and pick below were gathering simultaneously the layers of wealth that Nature had stored in her parlor and cellar for man.

I passed through Barnsley and Wakefield on this day's walk,—towns full of profitable industries and busy populations, and growing in both after the American impulse and expansion. If the good "Vicar of Wakefield" of the olden time could revisit the scene of his earthly experience, and look upon the old church of his ministry as it now appears, renovated from bottom to the top of its grand and lofty spire, he would not be entrapped again so easily into assent to the Greek apothegm of the swindler.

I lodged at a little village inn between Wakefield and Leeds, after a day of the most enjoyable walk that I had made. Never before, between sun and sun, had I passed over such a section of above-ground and under-ground industry and wealth. The next morning I continued northward, and noticed still more striking combinations of natural productions and human industries than on the preceding day. One small, rural area in which these were blended impressed me

greatly, and I stopped to photograph the scene on my mind. In a circle hardly a third of a mile in diameter, there was the heaviest crop of oats growing that I had yet seen in England; in another part of the same field there was a large brick-kiln; in another, an extensive quarry and machinery for sawing the stone into all sizes and shapes; then a furnace for casting iron, and lastly, a coal mine; and all these departments of labor and production were in full operation. It is quite possible that not one of the hundred laborers on and under this ten-acre patch ever thought it an extraordinary focus of production. Perhaps even the proprietors and managers of the five different enterprises worked on the small space had taken its rich and diversified fertilities as a matter of course, as we take the rain, light and heat of summer; but to a traveller "taking stock" of a country's resources, it could not but be a point of view exciting admiration. I left it behind me deeply impressed with the conviction that I had seen the most productive ten-acre field that could be found on the surface of the globe, counting in the variety and value of its surface and sub-surface crops.

I took tea with a friend in Leeds, remaining only an hour or two in that town, then pursuing my course northward. The wide world knows so much of Leeds that any notice that I could give of it might seem affected and presumptuous. It is to the Cloth-World what Rome is to the Catholic. Its Cloth Hall is the St. Peter's of Coat-and-trouserdom. Its rivers, streams and canals run black and blue with the stringent juices of all the woods and weeds of the world used in dyeing. The woods of all the continents come floating in here, like baled summer clouds of heaven. It is a city of magnipotent chimneys; and they stand thick and tall on the hills and in the valleys around, and puff their black breathings into the face and eyes of the sky above, baconising its countenance, and giving it no time to wash up and look sober, calm and clean, except a few hours on the sabbath. *The Leeds Mercury* is a power in the land, and everybody who reads the English

language in either hemisphere knows Edward Baines by name.

As I emerged from the great, busy town on the north, I passed by the estates and residences of its manufacturing aristocracy. The homes they have built and embellished should satisfy the tastes and ambitions of any hereditary nobility. They need only a little more age to make them rival many baronial establishments. It is interesting to see how the different classes of society are stepping into each other's shoes in going up into higher grades of social life. The merchant and manufacturing princes of England have not only reached but surpassed the conditions of wealth, taste and elegance which the hereditary peers of the realm occupied a century ago; while the latter have gone up to the rich and luxurious surroundings of kings and queens of that period. The upward movement has reached the very lowest strata of society. Not only have the small tradesmen and farmers ascended to the comfortable conditions of large merchants and landowners of one hundred years ago, but common day laborers are lifted upward by the general uprising. I should not wonder if all the damp, low cellarless cottages they now frequently inhabit should be swept away in less than fifty years and replaced by as comfortable buildings as the great middle class occupied in the childhood of the present generation.

I found comfortable quarters for the night in the little village of Bramhope, about five miles from Leeds. The next day I walked to Harrogate, passing through Otley and across the celebrated Wharf Vale. The scenery of this valley, as it opens upon you suddenly on descending from the south into Otley, is exceedingly beautiful; not so extensive as that of Belvoir Vale, but with all the features of the latter landscape compressed in a smaller space; like a portrait taken on a smaller scale. As you look off from the southern ridge or wall of the valley, you seem to stand on the cord of a segment of a circle, the radius of which touches the horizon at about five miles to the north. This crescent is filled with the most delicate lineaments of Nature's beauty. The opposite walls of the gallery slope upward from the meandering Wharf so gently and yet reach the blue ceiling of the sky so near, that all the paintings that panel them are vividly distinct to your eye, and you can group all their lights and shades in the compass of a single glance.

On the opposite side, half hidden and half revealed among the trees of an ample park, stands Farnley Hall, a historical residence of an old historical family. I had a letter of introduction to the present proprietor, Mr. Fawkes, who, I hope, will not deem it a disparagement to be called one of the Knights of the Shorthorns—a more extensive, useful, and cosmopolitan order than were the Knights of Rhodes or of Malta. Unfortunately for me, he was not at home; but his steward, a very intelligent, gentlemanly and genial man, took me over the establishment, and showed me all the stock that was stabled, mostly bulls of different ages. They were all of the best families of Shorthorn blood, and a better connoisseur of animal life than myself could not have enjoyed the sight of such well-made creatures more thoroughly than I did. The prince of the blood, in my estimation, was "Lord Cobham," a cream-colored bull, with which compared that famous animal in Greek mythology which played himself off as such an Adonis among the bovines, must have been a shabby, scraggy quadruped. Poor Europa! it would have been bad enough if she had been run away with by a "Lord Cobham." But the like of him did not live in her day.

After going through the housings for cattle, the steward took me to the Hall, a grand old mansion full of English history, especially of the Commonwealth period. Indeed, one large apartment was a museum of relics of that stirring and stormy time. There, against the antique, carved wainscoting, hung the great broad-brim of Oliver Cromwell, with a circumference nearly as large as an opened umbrella, heavy, coarse and grim. There hung a sword he wielded in the fiery rifts of battle. There was Fairfax's sword hanging by its side; and

his famous war-drum lay beneath. Its leather lungs, that once shouted the charge, were now still and frowsy, with no martial speech left in them.

Mr. Fawkes owns about 15,000 acres of land, including most of the valley of Otley, and extending back almost to Harrogate. He farms about 450 acres, but grows no wheat. Indeed, I did not see a field of it in a circle of five miles' diameter.

I reached Harrogate in the dusk of the evening, and found the town alive with people mostly in the streets. It is a snug and cozy little Saratoga among the hills of Yorkshire, away from the smoke, soot and savor of the great manufacturing centres. It is a favorite resort for a mild class of invalids, and of persons who need the medicine of pure air and gentle exercise, blended with the quiet tonics of cheery mirth and recreation. Superadded to all these stimulants, there is a mineral spring at which the visitors, young and old, drink most voluminously. I went down to it in the morning before breakfast, and found it thronged by a multitude of men, women and children, who drank off great goblets of it with astonishing faith and facility. The rotunda was so filled with the fumes of sulphur that I found it more easy to inhale than to imbibe, and preferred to satisfy that sense as to the merits of the water.

The next day I reached the brave old city of Ripon. On the way I stopped an hour or two at Ripley and visited the castle. The building itself is a good specimen of the baronial hall of the olden time. But the gardens and grounds constitute its distinguishing feature. I never saw before such an exquisite arrangement of flowers, even at Chatsworth or the Kew Gardens. All forms imaginable were produced by them. The most extensive and elaborate combination was a row of flower sofas reaching around the garden. Each was from 20 to 30 feet in length. The seat was wrought in geraniums of every tint, all grown to an even, compact surface, presenting figures as diversified as the alternating hues could produce. The back was worked in taller flowers, pre-

senting the same evenness of line and surface. On entering the garden gate and catching the first sight of these beautiful structures, you take them for veritable sofas, as perfectly wrought as anything was ever done in Berlin wool.

Ripon is an interesting little city, with a fact-roll of history reaching back into the dimmest centuries of the land. It has run the gauntlet of all the Saxon, Danish, Scotch and Norman raids and regimes. It was burnt once or twice by each of these races in the struggle for supremacy. But with a plucky tenacity of life, it arose successively out of its own ashes and spread its phœnix wings to a new and vigorous vitality. A venerable cathedral looks down upon it with a motherly face. Unique old buildings, with half their centuries unrecorded and lost in oblivion, stand to this day in good repair, as the homes of happy children, who play at marbles and the last sports of the day just as if they were born in houses only a year older than themselves. Institutions and customs older than the cathedral are kept up with a filial faith in their virtue. One of the most interesting of these, I believe, was established by the Saxon Edgar or Alfred—it matters not which; they were only a century or two apart, and that space is but a trifling circumstance in the history of this old country. One of these kings appointed an officer called a "wakeman" for the town. He must originally have been a kind of secular beadle of the community, or a curfew constable, to see the whole population well a-bed in good season. One of his duties consisted in blowing a horn every night at nine o'clock as a signal to turn in. But a remarkable consideration was attached to faithful compliance with this summons. If any house or shop was robbed before sunrise, a tax was levied upon every inhabitant, of 4d. if his house had one outer door, and of 8d. if it had two. This tax was to compensate the sufferer for his loss, and also to put the whole community under bonds to keep the peace and to feel responsible for the safety of each other's property. Thus it not only acted as a great mutual insurance company of which every householder was a member, but it made him, as it were, a special constable against burglary. This old Saxon institution is in full life and vigor to-day. The wakeman is still the highest secular official of the town. For a thousand consecutive years the wakeman's toot-horn has been blown at night over the successive generations of the little cathedral city. This is an interesting fact, full of promise. No American could fail to admire this conservatism who appreciates national individuality. No one, at heart, could more highly esteem these salient traits of a people's character. And here I may as well put in a few thoughts on this subject as at any stage of my walk.

Good-natured reader, are you a man of sensitive perceptions as to the proprieties and dignities of dress and deportment which should characterise some great historical personage whose name you have held in profound veneration all your life long? Now, in the wayward drift of your imagination among the freaks of modern fashion, did it ever dare to present before your eyes St. Paul in strapped pantaloons, figured velvet vest, swallow-tailed coat, stove-pipe hat, and a cockney glass at his eye? Did your fancy, in its wildest fictions, ever pass such an image across the speculum of your mental vision?

Gentle reader, "in maiden meditation, fancy free," did a dreamy thought of yours ever stray through the histories of your sex and its modes of dress and adornment, and so blend or transpose them as to present to you, in a sudden flash of the imagination, the Virgin Mary dressed like the Empress Eugenie? Readers both, did not that fancy trouble you, as if an unholy thought had fallen into the soul? Well, a thought like that must trouble the American when his fancy passes before his mind's eye the image of Old England *Americanised*. And a faculty more serious and trusty than fancy will present this transformation to him, day by day, as he visits the great centres of the nation's life and industry. In London, Manchester, Liverpool, and all the most busy and prosperous commercial and manufacturing towns, he will see that England

is becoming Americanised shockingly fast. In all these populous places it is losing the old individuality that once distinguished the grandfatherland of fifty millions who now speak its language beyond the sea. Look at London! look at the miles of three and four story houses under the mason's hands, now running out in every direction from the city. Will you see a single feature of the Old England of our common memories in them? No, not one! no more than in a modern English dress-coat, or in one of the iron rails of the British Great Western, or of the Illinois Central. It is doubtful if there will be anything of England left in London at the end of the next fifty years, unless it be the fog and the Lord Mayor's Show. Already the radicals are crying out against both of these institutions, which are merely local, by the way. The tailor's shears, the mason's trowel, and the carpenter's edge-tools are evening everything in Christendom to one dead level of uniformity. The railroads and telegraphs are all working to the same end. All these agencies of modern civilization at first lay their innovating hands upon large cities or commercial centres. Thence they work outward slowly and transform the appearance and habits of the country. The transformations I have noticed in England since 1846 are wonderful, utilitarian, and productive of absolute and rigid comfort to the people; still, I must confess, they inspire in me a sentiment akin to that which our village fathers experienced when the old church in which they worshipped from childhood was pulled down to make room for a better one.

To every American, sympathising with these sentiments, it must be interesting to visit such a rural little city as Ripon, and find populations that cling with reverence and affection to the old Saxon institutions of Alfred. It will make him feel that he stands in the unbroken lineage of the centuries, to hear the wakeman's horn, and to know that it has been blown, spring, summer, autumn and winter, in all weathers, in weal and in woe, for a thousand years. As Old England is driven farther and

farther back from London, Manchester, Liverpool, and other great improving towns, she will find refuge and residence in these retired country villages. Here she will wear longest and last the features in which she was engraven on the minds of all the millions who call her mother beyond the sea.

The next day I visited the celebrated Fountain Abbey in Studley Park,—a grand relic of antiquity, framed with silver and emerald work of lakelets, lawns, shrubberies and trees as beautifully arranged as art, taste and wealth could set them. The old abbey is a majestic ruin which fills one with wonder as he looks up at its broken arches and towers and sees the dimensions marked by the pedestals or foot-prints of its templed columns. It stands rather in a narrow glen than in a valley, and was commenced, it is supposed, about 1130. The yew-trees under which the monks bivouacked while at work upon the magnificent edifice, are still standing, bearing leaves as large and green as those that covered the enthusiastic architects of that early time. In the height of its prosperity and power, the lands of the abbey embraced over 72,000 acres. The Park enclosing this great monument of an earlier age contains 250 acres, and is really an earthly elysium of beauty. It was comforting to learn that it was laid out so late as 1720, and that all the noble trees that filled it had grown to their present grandeur within the intervening period. Here I saw for the first time in England our hard-maple. It was a spindling thing, looking as if it had suffered much from fever and ague or rheumatism; but it was pleasant to see it admitted into a larger fellowship of trees than our New England soil ever bore. On a green, lawn-faced slope, at the turning of the principal walk, there was a little tree a few feet high enclosed in by a circular wire fence. It was planted by the Princess of Wales on a visit of the royal pair to Studley soon after their marriage. The fair Dane left her card in this way to the old Abbey, which began to rise upon its foundations soon after the stalwart Danish sovereign of England fell at the Battle of Hastings.

Will any one of her posterity ever bear his name and sit upon the throne he vacated for that bloody grave? No! She will remember a better name at the font. The day and the name of the Harolds, Williams, Henrys, Charles's, and Georges are over and gone forever. ALBERT THE GOOD has estopped that succession; and England, doubtless, for centuries to come, will wear that name and its memories in her crown.

After spending a few hours at Studley Park, I returned to Ripon and went on to Thirsk, where I spent the Sabbath with a Friend. The next day he drove me over to Rievaulx Abbey, which was the mother of Fountain Abbey. On the way to it we passed the ruins of another of these grand structures of that religious age, called Byland Abbey, where Robert Bruce came within an ace of capturing King Edward on his retreat from Scotland, after the Battle of Bannockburn.

One of the objects of this excursion was to visit the establishment of Lord Faversham, near Helmsley, who is one of the most scientific and successful stock-raisers, of the Shorthorn blood, in England, and to whom I had a note of introduction. But he, too, was not at home, which I much regretted, as I was desirous of seeing one of the peers of the realm who enter into this culture of animal life with so much personal interest and assiduity. His manager, however, was very affable and attentive, ready and pleased to give any information desired upon different points. He showed us a splendid set of animals. Indeed, I had never seen a herd to equal it. There were several bulls of different ages with a perfection of form truly admirable. Some of them had already drawn first prizes at different shows. Several noble specimens of this celebrated herd have been sold to stock-raisers in America, Australia and in continental countries. The most perfect of all the well-made animals on the establishment, according to my untrained perceptions of symmetry, was a milk-white cow, called "The Lady in White," three years old. She and Mr. Fawkes' "Lord Cobham" should be shown together. I doubt if a

better mated pair could be found in England. There was a large number of cows feeding in the park which would command admiration at any exhibition of stock. Lord Faversham's famous "Skyrocket" ended his days with much *eclat*. When getting into years, and into monstrous obesity, he was presented as a contribution to the Lancashire Relief Fund. Before passing into the butcher's hands, he was exhibited in Leeds, and realised about £200 as a show. Thus as a curiosity first, and as a small mountain of fat beef afterward, he proved a generous gift to the suffering operatives in the manufacturing districts.

Passing through the park gate, we entered upon a lawn esplanade looking down upon the ruins of Rievaulx Abbey. This broad terrace extended for apparently a half of a mile, and was as finely carpeted piece of ground as you will find in England. No hair of horse or dog groomed and brushed with the nicest care, and soft and shining with the healthiest vitality, could surpass in delicacy and life of surface the grass coverlet of this long terrace, from which you looked down upon that grand monument of twelfth-century architecture half veiled among the trees of the glen. This was one of the oldest abbeys in the north of England, and the mother of several of them. Some of its walls are still as entire and perfect as those of Tintern, on the Wye. It was founded by the monks of the St. Bernard order, in 1131, according to the historical record. Really those black-cowled masons and carvers must have given the enthusiasm and genius of the early painters of the Virgin to these magnificent structures. I will not go into the subject at large here, leaving it to form an entire chapter, when I have seen most of the old abbeys of the country. In looking up at their walls, arches and columns, one marvels to see the most delicate and elaborate vine and flower-work of the carver's chisel apparently as perfect as when it engraved the last line; and this, too, in face of the frosts and beating storms of six hundred years. The largest ivy I ever saw buttressed one of the windowed walls with ten thousand cross-

folded fingers and foliage of vivid green piled thick and high upon the teethmarks of time. The trunk was a full foot through at the butt. A few years ago a large mound was uncovered near the ruin, and found to be composed of cinders, showing incontestably that the monks had worked iron ore very extensively, thus teaching the common people that art as well as agriculture. These cinders have been used very largely in repairing the roads for a considerable distance around.

On returning to Thirsk over the Hambleton range of hills, we crossed thousands of acres of moor-land covered with heather in full bloom, looking like a purple sea. It was a splendid sight. My friend, who was an artist, stopped for a while to sketch one or two views of the scene. As we proceeded, we saw several green and golden fields impinging upon this florid waste, serving to illustrate what might be done with the vast tracts of land in England and Scotland now bristling with this thick and prickly vegetation. The heatherland over which we were passing was utilised in a rather singular manner. It yielded pasturage to two sets of industrials—sheep and bees. As the heather blossom is thought to impart a peculiarly pleasant flavor to honey, I was told many bee-stock-raisers of Lincolnshire brought their hives to this section to pasture them for a season on this purple prairie.

The westward view from the precipitous heights of the Hambleton ridge is one of the most beautiful and extensive you will find in England, well worth a special journey to see it. The declining sun was flooding the great basin with the day's last, best smile, filling it to the golden rim of the horizon with a soft light in which lay a landscape of thirty miles' depth, embracing full fifty villages and hamlets, parks, plantations and groves, all looking "like emeralds chased in gold." On the whole, I am inclined to think many tourists would regard this view as even superior to that of Belvoir Vale. It might be justly placed between that and Wharf Vale.

A London gentleman produced a most unique picture on the forehead of one of these hills, which may be seen at a great distance. In the first place, he had a smooth, lawn-like surface prepared on the steep slope. Then he cut out the form of a horse in the green turf, sowing the whole contour of the animal with lime. This brought out in such bold relief the body and limbs, that, at several miles distance, you seem to see a colossal white horse standing on his four legs, perfect in form and feature, even to ear and nostril. The symmetry is perfect, although the body, head, legs and tail cover a space of *four* acres!

The next day I took staff for Northallerton, reaching that town about the middle of the afternoon. Passed through a highly cultivated district, and saw, for the first time, several reaping machines at work in the fields. I was struck at the manner in which they were used. I have noticed a peculiarity in reaping in this section which must appear singular to an American. The men cut inward instead of outward, as with us. And these machines were following the same rule! As they went around the field, they were followed or rather met by men and women, each with an allotted beat, who rushed in behind and gathered up the fallen from the standing grain so as to make a clear path for the next round. There seemed to be no reason for this singular and awkward practice, except the adhesion to an old custom of reaping. The grain was not very stout, nor was it lodged.

From Northallerton I hastened on to Newcastle-upon-Tyne in order to attend, for the first time in my life, the meetings of the British Association. I reached that town on the 25th of August, and remained there a week, enjoying one of the greatest treats that ever fell to my lot. I will reserve a brief description of it for a separate chapter at the end of this volume, if my Notes on other matters do not crowd it out.

CHAPTER XVI.

HEXHAM—THE NORTH TYNE— BORDER-LAND AND ITS SUGGESTIONS—HAWICK—TEVIOTDALE— BIRTH-PLACE OF LEYDEN—MELROSE AND DRYBURGH ABBEYS— ABBOTSFORD: SIR WALTER SCOTT; HOMAGE TO HIS GENIUS—THE FERRY AND THE OAR-GIRL—NEW FARM STEDDINGS—SCENERY OF THE TWEED VALLEY—EDINBURGH AND ITS CHARACTERISTICS.

On Thursday, Sept. 3rd, I left Newcastle, and proceeded first westward to the old town of Hexham, with the view of taking a more central route into Scotland. Here, too, are the ruins of one of the most ancient of the abbeys. The parish church wears the wrinkles of as many centuries as the oldest in the land. Indeed, the town is full of antiquities of different dates and races,—Roman, Scotch, Saxon, Danish and Norman. They all left the marks of their glaived hands upon it.

From Hexham I faced northward and followed the North Tyne up through a very picturesque and romantic valley, thickly wooded and studded with baronial mansions, parks, castles and residences of gentry, with comfortable farm-houses looking sunny and cheerful on the green hill slopes and on the quiet banks of the river. I saw fields of wheat quite green, looking as if they needed another month's sun to fit them for harvesting. Lodged in a little village about eight miles from Hexham. The next day walked on to the little hamlet of Fallstones, a distance of about twenty miles. As I ascended the valley, the scene changed rapidly. The river dwindled to a narrow stream. The hills that walled it in on either side grew higher and balder, and the clouds lay cold and dank upon their bleak and sullen brows. The hamlets edged in here and there grew thinner, smaller and shabbier. The road was barred and gated about once in a mile, to keep cattle and sheep from wandering; there being no fences nor hedges running parallel with it. In a word, the premonitory symptoms of a bare border-land thickened at every turn.

Another day brought me into the midst of a wild region, which might be called No-man's-land; although most of it belongs to the Duke of Northumberland. It is all in the solitary grandeur of heather-haired hills, which tinge, with their purple flush, the huge, black-winged clouds that alight upon them. Only here and there a shepherd's cottage is to be seen half way up the heights, or sheltering itself in a clump of trees in glen or gorge, like a benighted traveller bivouacking for a night in a desert. Sheep, of the Cheviot breed mostly, are nearly the sole inhabitants and industrials of this mountainous waste. They climb to the highest peaks and bring down the white wealth of their wool to man. It was pleasant to see them like walking mites, flecking the dark brows of the mountains. They made a picture; they made a *tableau vivant* of the same illustration as Landseer's lamb looking into the grass-covered cannon's mouth.

This is the Border-land! Here the fiercest antagonisms of hostile nationalities met in deadly conflict. Fire and blood, rapine and wrath blackened and reddened and ravaged for centuries across this bleak territory. Robber-chieftains and knighted free-booters carried on their guerilla raids backward and forward, under the counterfeited banner of patriotism. Scotch and English armies led by kings marched and counter-marched over this sombre boundary. Never before was there one apparently more insoluble as a barrier between two peoples. Never before in Christendom was there one that required a longer space of time to melt. Never before did the fusing of two nationalities encounter more fierce and prolonged opposition. Did ever patriotism pour out a swifter and deeper tide of chivalrous sentiment against merging one in another?—against uniting two thrones and two peoples in one? Did patriotism ever fight bloodier battles to prevent such a union, or cling to local sovereignty with a more desperate hold?

This is the Border-land! Look up the purpled steeps of these heathered hills. The white lambs are looking, with their soft, meek eyes, into the grass-choked mouths of the rusty and dismantled cannon of the war of nationalities between England and Scotland. The deed has been consummated. The valor and patriotism of Wallace and Bruce could not prevent it. The sheep of English and Scotch shepherds feed side by side on these mountain heights, in spite of Stirling and Bannockburn, of Flodden and Falkirk. The Iron Horse, bearing the blended arms of the two realms on his shield, walks over those battle-fields by night and day, treading their memories deeper and deeper in the dust. The lambs are playing in the sun on the boundary line of the two dominions. Does a Scot of to-day love his native land less than the Campbell clansman or clan-chief in Bruce's time? Not a whit. He carries a heartful of its choicest memories with him into all countries of his sojourning. But there is a larger sentiment that includes all these filial feelings towards his motherland, while it draws additional warmth and strength from them. It is the sentiment of Imperial Nationality; the feeling of a Briton, that does not extinguish nor absorb, nor compete with, the Scot in his heart;— the feeling that he is a political constituent of a mighty nation, whose feet stand upon all the continents of the earth, while it holds the best islands of the sea in its hands;—the feeling with which he says *We* with all the millions of a dominion on which the sun never sets, and *Our*, when he speaks of its grand and common histories, its hopes, prospects, progress, power and aspirations.

There was a Border-land, dark and bloody, between Saxon England and Celtic Wales. For centuries the red footmarks of savage conflict scarred and covered its wild waste. Never before did so small a people make so stout, and desperate and protracted struggle for local independence and isolation. Never did one produce a more strong-hearted and blind-eyed patriotism, or patriotism more poets to thrill the listeners to their lays with the intoxicating fanaticism of a national sentiment. On that Border-land the white lambs now lie in the sun. The Welsh sentiment is as strong as ever in the Snowdon shepherd, and he

may not speak a dozen words of the English tongue. But the Briton lives in his breast. The feeling of its great meaning surrounds and illumines the inner circles of his local attachment. He may never have seen a map of the Globe, and never have been outside the wall of the Welsh mountains; but he knows, without geography, who and what Queen Victoria is among the earth's sovereigns, and the length and breadth of her sceptre's reach and rule around the world.

There was a Border-land between Britain and Ireland, blackened and scarred by more burning antagonisms than those that once divided the larger island. The record of several consecutive centuries is graven deep in it by the brand and bayonet, and by the more incisive teeth-marks of hate. The slumbering antipathies of race and religion even now crop out here and there, over the unfused boundary, in hissing tongues of flame. The Briton and the Celt are still struggling for the precedence in the Irishman's breast; but it is not a war of extermination. His ardent nature is given to martial memories, and all the battles he boasts of are British battles, in which he or his father played the hero number one. The history of independent Ireland is poor and thin; still he holds it back in his heart, and hesitates to link it with the great annals of the "Saxon" realm, and thus make of both one grand and glorious record, present and future. He cannot yet make up his mind to say *We* with all the other English-speaking millions of the empire, as the Scotsman and Welshman have learned and loved to say it. He cannot as yet say *Our* with them with such a sentiment of joint-interest, when the histories, hopes, expansion and capacities of that empire unroll their vista before him. But the rains and the dews of a milder century are falling upon this Border-land. The lava of spent volcanoes that covered it is taking soil and seed of green vegetation. The white lambs shall yet lie on it in the sun.

What a volume might be filled with the succinctest history of the Border-lands of Christendom! France was intersected with them for centuries. Seem-

ingly they were as implacable and obdurate as any that ever divided the British isle. Local patriotism wrote poetry and shed blood voluminously to prevent the fusion of these old landmarks of pigmy nationalities. It took nearly a thousand years to complete the blending; to make the *we* and the *our* of one great consolidated empire the largest political sentiment of the men of Normandy, Burgundy or Navarre. Long and fierce, and seemingly endless was the struggle; but at last, on all those old obstinate boundaries of hostile principalities, the white lambs lay in the sun.

There are Border-lands now in the south and east of Europe foaming and seething with the same antagonisms of race and language; and Christendom is tremulous with their emotion. It is the same old struggle over again; and yet ninety-nine in a hundred of intelligent and reading people, with the history of British and French Border-lands before them, seem to think that a new and strange thing has happened under the sun. Full that proportion of our English-speaking race, in both hemispheres, closing the volume of its own annals, have made up their minds to the belief that these Border-lands between German and Magyar, Teuton and Latin, Russ and Pole, bristle with antagonisms the like of which never were subdued, and never ought to be subdued by human means or motives. To them, naturally, the half century of this hissing and seething, insurrection and repression, is longer than the five hundred years and more it took to fuse into one the nationalities of England and Wales. What a point of space is a century midway between the ninth and nineteenth! Few are long-sighted enough in historic vision to touch that point with a cambric needle. It may seem unfeeling to say it or think it; still it is as true as the plainest history of the last millenium. There is a patriotism that looks at the future through a gimlet hole, and sees in it but a single star. That patriotism is a natural, and most popular sentiment. It was strong in the Welshman's breast a thousand years ago, and in the Scotsman's half that distance back in the past. But it is a patriotism that has its day and its rule; then both its eyes are opened, and it looks upon the firmament of the future broadside on, and sees a constellation where it once saw and half worshipped a solitary star. Better to be the part of a great WHOLE than the whole of a little *nothing*.

These continental Border-lands may see the face of their future history in the mirror of England's annals. They are quaking now with the impetuous emotions of local nationality. They are blackened and scarred in the contest for the Welsh and Scotch independence of centuries agone. But over those boundary wastes the grass shall yet grow soft, fair and green, and there, too, the white lambs shall lie in the sun.

My walk lay over the most inhospitable and unpeopled section I ever saw. Calling at a station on the railway that passes through it, I was told by the master that the nearest church or chapel was sixteen miles in one direction, and over twenty in another. It is doubtful if so large a churchless space could be found in Iowa or even Kansas. I was glad to reach Hawick, a good, solid town but a little way inside of the Scottish border, where I spent the sabbath and the following Monday. This was a rallying and sallying point in the old Border Wars, and was inundated two or three times by the flux and reflux of this conflict, having been burnt twice, and put under the ordeal of other calamities brought upon it when free-booting was both the business, occupation and pastime of knighted chieftains and their clansmen. It is now a thrifty, manufacturing town, lying in the trough of the sea, or of the lofty hills that resemble waves hardened to earth in their crests. Just opposite the Temperance Inn in which I had my quarters, was the Tower Hotel, once a palatial mansion of the Buccleuchs. There the Duchess of Monmouth used to hold her drawing-rooms in an apartment which many a New England journeyman mechanic would hardly think ample and comfortable enough for his parlor. There is a curious conical mound in the town, called the Moat-hill, which looks like a great,

green carbuncle. It is thought by some to be a Druidical monument, but is quite involved in a mystery which no one has satisfactorily solved. It is strange that no persistent and successful effort has been made to let day-light through it. Some workmen a long time ago undertook to perforate it, but were frightened away by a thunder-storm, which they seemed to take as a reproof and threatened punishment for their profanity. The great business of Hawick is the manufacture of a woollen fabric called *Tweeds*. It came to this name in a singular way. The clerk of the factory made out an invoice of the first lot to a London house under the name of *Twilled* goods. The London man read it *Tweeds*, instead of Twilled, and ever since they have gone by that title. As Sir Walter Scott was at that time making the name "Tweed" illustrious, the mistake was a very lucrative one to the manufacturers of the article. Here, too, in this border town commences the chain of birthplaces of eminent men, who have honored Scotland with their lives and history. Here was born James Wilson, once the editor of *The Economist*, who worked his way up, through intermediate positions of public honor and trust, to that of Finance Minister for India, and died at the meridian of his manhood in that country of dearly-bought distinctions.

On Tuesday, Sept. 8th, I commenced my walk northward from this threshold town of Scotland. Followed down the Teviot to Denholm, the birth-place of the celebrated poet and linguist, Dr. John Leyden, another victim who offered himself a sacrifice to the costly honors and emoluments of East Indian official life. One great thought fired his soul in all the perils and privations of that deadly climate. It was to ascend one niche higher in knowledge of oriental tongues than Sir William Jones. He labored to this end with a desperate assiduity that perhaps was never surpassed or even equalled. He died hugging the conviction that he had attained it. This little village was his birthplace. Here he wrote his first rhymes, and wooed and won the first inspirations of the muse. His heart, as its last pulses

grew weaker and slower, in that far-off heathen land, took on its child-thoughts again and its child-memories; and his last words were about this little, rural hamlet where he was born. A beautiful monument has been erected to his memory in the centre of the large common around which the village is built. On each of the four sides of the monument there is a tribute to his name and worth; one from Sir Walter Scott, and one taken from his own poems, entitled "Scenes of my Infancy," a touching appeal to his old friends and neighbors to hold him in kind remembrance.

All this section is as fertile as it can be in the sceneries and historical associations favorable for inspiring a strong-hearted love of country, and for the development of the poetry of romantic patriotism. It was pleasant to emerge from the dark, cold, barren border-land, from the uncivilized mountains, standing sullen in the wild, shaggy *chevelure* of nature, and to walk again between towering hills dressed in the best toilet of human industry, crowned with golden wheatfields, and zoned with broad girdles of the greenest vegetation. It is when these contrasts are suddenly and closely brought within the same vista that one sees and feels how the Creator has honored the labor of human hands, and lifted it up into partnership with His omnipotences in chronicling the consecutive centuries of the earth in illuminated capitals of this joint handwriting. It is a grand and impressive sight— one of those dark-browed hills of the Border-land, bearded to its rock-ridged forehead with such bush-bristles and haired with matted heather. In nature it is what a painted Indian squaw in her blanket, eagle feathers and moccasins, is in the world of humanity. We look upon both with a species of admiration, as contrasts with objects whose worth is measured by the comparison. The Empress Eugenie and the Princess of Wales, and wives and sisters lovelier still to the circles of humble life, look more beautiful and graceful when the eye turns to them from a glance at the best-looking squaw of the North American wilds. And so looked the well-dressed hills on each side of the Teviot, compared with the uncultured and stunted mountains among which I had so recently walked.

Ascending from Teviotdale, I passed the Earl of Minto's seat, a large and modern-looking mansion, surrounded with beautiful grounds and noble trees, and commanding a grand and picturesque view of valley and mountain from an excellent point of observation. As soon as I lost sight of Teviotdale another grand vista of golden and purpled hills and rich valleys burst upon my sight as suddenly as theatrical sceneries are shifted on the stage. Dined in a little, rural, unpoetical village bearing the name of Lilliesleaf. Resuming my walk, I soon came in sight of the grand valley of the Tweed, a great basin of natural beauty, holding, as it were, Scotland's "apples of gold in pictures of silver." Every step commanded some new feature of interest. Here on the left arose to the still, blue bosom of the sky the three great Eildon Hills, with their heads crowned with heather as with an emerald diadem. The sun is low, and the far-off village in the valley shows dimly between the daylight and darkness. There is the shadow of a broken edifice, broken but grand, that arises out of the midst of the low houses. A little farther on, arches, and the stone vein-work of glassless windows, and ivy-netted towers come out more distinctly. I recognise them at the next furlong. They stand thus in pictures hung up in the parlors of thousands of common homes in America, Australia and India. They are the ruins of Melrose Abbey. Here is the original of the picture. I see it at last, as thousands of Americans have seen it before. In history and association it is to them the Westminster Abbey of Scotland, but in ruin. It looks natural, though not at first glance what one expected. The familiar engraving does not give us the real flesh and blood of the antiquity, or the complexion of the stone; but it does not exaggerate the exquisite symmetries and artistic genius of the structure. These truly inspire one with wonder. They are all that pen and pencil have described them. The great win-

dow, which is the most salient feature in the common picture, is a magnificent piece of work in stone, twenty-four feet in height and sixteen in breadth. It is all in the elm-tree order of architecture. The old monks belonged to that school, and they wrought out branches, leaves and leaf-veins, and framed the lacework of their chisels with colored glass most exquisitely.

Melrose Abbey was the eldest daughter, I believe, of Rievaulx Abbey, in Yorkshire, which has already been noticed; a year or two older in its foundation than Fountain Abbey, in Studley Park. The fecundity with which these ecclesiastical buildings multiplied and replenished England and Scotland is a marvel, considering the age in which they were erected and the small population and the poverty of the country. But something on this aspect of the subject hereafter. Here lie the ashes of Scottish kings, abbots and knights whose names figured conspicuously in the history of public and private wars which cover such a space of the country's life as an independent nation. The Douglas family especially with several of its branches found a resting-place for their dust within these walls. Built and rebuilt, burnt and reburnt, mutilated, dismembered, consecrated and desecrated, make up the history of this celebrated edifice, and that of its like, from Land's End to John O'Groat's. It is a slight but a very appreciable mitigation of these destructive acts that it was ruined *artistically*; just as some enthusiastic castle and abbey-painter would have suggested.

Although I spent the night at Melrose, it was a dark and cloudy one, so that I could not see the abbey by moonlight—a view so much prized and celebrated. The next day I literally walked from morning till evening among the tombstones of antiquity and monuments of Scotch history invested with an interest which will never wane. In the first place, I went down the Tweed a few miles and crossed it in a ferry-boat to see Dryburgh Abbey. Here, embowered among the trees in a silver curve of the river, stands this grand monument of one of the most remarkable ages of the

world. Within an hour's walk from Melrose, and four or five years only after the completion of that edifice, the foundations of this were laid. It is astonishing. We will not dwell upon it now, but make a separate chapter on it when I have seen most of the other ruins of the kind in the kingdom. The French are given to the habit of festooning the monuments and graves of their relatives and friends with *immortelles*. Nature has hung one of hers to Dryburgh Abbey. It is a yew-tree opposite the door by which you enter the ruins. The year-rings of its trunk register all the centuries that the stones of the oldest wall have stood imbedded one upon the other. The tree is still green, putting forth its leaf in its season. But there is an *immortelle* hung to these dark, crumbling walls that shall outlive the greenest trees now growing on earth. Here, in a little vaulted chapel, or rather a deep niche in the wall, lie the remains of Sir Walter Scott, his wife and the brilliant Lockhart. How many thousands of all lands where the English language is spoken will come and stand here in mute and pensive communion before the iron gate of this family tomb and look through the bars upon this group of simply-lettered stones!

From Dryburgh I walked back to Melrose on the east side of the Tweed. Lost the footpath, and for two hours clambered up and down the precipitous cliffs that rise high and abrupt from the river. In many places the zig-zag path was cut into the rock, hardly a foot in breadth, overhanging a precipice which a person of weak nerves could hardly face with composure. At last got out of these dark fastnesses and ascended a range of lofty hills where I found a good carriage road. This elevation commanded the most magnificent view that I ever saw in Scotland, excepting, perhaps, the one from Stirling Castle only for the feature which the Forth supplies. It was truly beautiful beyond description, and it would be useless for me to attempt one.

After dinner in Melrose, I resumed my walk northward and came suddenly upon Abbotsford. Indeed, I should have missed it, had I not noticed a wooden gate open on the roadside, with some directions upon it for those wishing to visit the house. As it stands low down towards the river, and as all the space above it to the road is covered with trees and shrubbery, it is entirely hidden from view in that direction. The descent to the house is rather steep and long. And here it is!—Abbotsford! It is the photograph of Sir Walter Scott. It is brim full of him and his histories. No author's pen ever gave such an individuality to a human home. It is all the coinage of thoughts that have flooded the hemispheres. Pages of living literature built up all these lofty walls, bent these arches, panelled these ceilings, and filled the whole edifice with these mementoes of the men and ages gone. Every one of these hewn stones cost a paragraph; that carved and gilded crest, a column's length of thinking done on paper. It must be true that pure, unaided literary labor never built before a mansion of this magnitude and filled it with such treasures of art and history. This will forever make it and the pictures of it a monument of peculiar interest. I have said that it is brim full of the author. It is equally full of all he wrote about; full of the interesting topographs of Scotland's history, back to the twilight ages; full inside and out, and in the very garden and stable walls. The studio of an artist was never fuller of models of human or animal heads, or of counterfeit duplicates of Nature's handiwork, than Sir Walter's mansion is of things his pen painted on in the long life of its inspirations. The very porchway that leads into the house is hung with petrified staghorns, doubtless dug up in Scottish bogs, and illustrating a page of the natural history of the country in some prehistoric century. The halls are panelled with Scotland,—with carvings in oak from the old palace of Dunfermline. Coats of arms of the celebrated Border chieftains are arrayed in line around the walls. The armoury is a miniature arsenal of all arms ever wielded since the time of the Druids. And a history attaches to nearly every one of the weapons. History hangs its webwork everywhere.

It is built, high and low, into the face of the outside walls. Quaint, old, carved stones from abbey and castle ruins, arms, devices and inscriptions are all here presented to the eye like the printed page of an open volume. Among the interesting relics are a chair made from the rafters of the house in which Wallace was betrayed, Rob Roy's pistol, and the key of the old Tolbooth of Edinburgh.

I was conducted through the rooms opened to visitors by a very gentlemanly-looking man, who might be taken for an author himself, from his intellectual appearance and conversation. The library is the largest of all the apartments—fifty feet by sixty. Nor is it too large for the collection of books it contains, which numbers about 20,000 volumes, many of them very rare and valuable. But the soul-centre of the building to me was the *study*, opening into the library. There is the small writing-table, and there is the plain armchair in which he sat by it and worked out those creations of fancy which have excited such interest through the world. That square foot over against this chair, where his paper lay, is the focus, the point of incidence and reflection, of thoughts that pencilled outward, like sun-rays, until their illumination reached the antipodes,—thoughts that brought a pleasant shining to the sun-burnt face of the Australian shepherd as he watched his flock at noon from under the shadow of a stunted tree; thoughts which made a cheery fellowship at night for the Hudson Bay hunter, in his snow-buried cabin on the Saskatchiwine. The books of this little inner library were the bodyguard of his genius, chosen to be nearest him in the outsallyings of his imagination. Here is a little conversational closet, with a window in it to let in the leaf-sifted light and air—a small recess large enough for a couple of chairs or so, which he called a "*Speak-a-bit*." Here is something so near his personality that it almost startles you like a sudden apparition of himself. It is a glass case containing the clothes he last wore on earth,—the large-buttoned, blue coat, the plaid trousers, the broad-

brimmed hat, and heavy, thick-soled shoes which he had on when he came in from his last walk to lay himself down and die.

On signing my name in the register, I was affected at a coincidence which conveyed a tribute of respect to the memory of the great author of striking significance, while it recorded the painful catastrophe which has broken over upon the American Republic. It was a sad sight to me to see the profane and suicidal antagonisms which have rent it in twain brought to the shrine of this great memory and graven upon its sacred tablet as it were with the murdering dagger's point. New and bad initials! The father and patriot Washington would have wept tears of blood to have read them here,—to have read them anywhere, bearing such deplorable meaning. They were U.S.A. and C.S.A., as it were chasing each other up and down the pages of the visitors' register. Sad, sad was the sight—sadder, in a certain sense, than the smoke-wreaths of the Tuscarora and Alabama ploughing the broad ocean with their keels. U.S.A. and C.S.A.! What initials for Americans to write, with the precious memories of a common history and a common weal still held to their hearts—to write here or anywhere! What a riving and a ruin do those letters record! Still they brought in their severed hands a common homage-gift to the memory of the Writer of Abbotsford. If they represented the dissolution of a great political fabric, in which they once gloried with equal pride, they meant union here—a oneness indissoluble in admiration for a great genius whose memory can no more be localised to a nation than the interest of his works.

American names, both of the North and South, may be found on almost every page of the register. I wrote mine next to that of a gentleman from Worcester, Mass., my old place of residence, who only left an hour before my arrival. Abbotsford and Stratford-upon-Avon are points to which our countrymen converge in their travels in this country; and you will find more of their signatures in the registry of these two haloed homesteads of genius than anywhere else in Europe.

The valley of the Tweed in this section is all an artist would delight in as a surrounding of such histories. The hills are lofty, declining into gorges or dells at different angles with the river, which they wall in precipitously with their wooded sides in many places. They are mostly cultivated to the top, and now in harvest many of them were crowned with stooked sheaves of wheat, each looking in the distance like Nature with her golden curls done up in paper, dressing for the harvest-home of the season. Some of them wore belts and gores of turnip foliage of different *nuances* of green luxuriance, combining with every conceivable shade and alternation of vegetable coloring. Indeed, as already intimated, the view from the eminence almost overhanging the little sequestered peninsula on which Old Melrose stood twelve centuries ago, is indescribably beautiful, and well worth a long journey to see, disconnected from its historical associations. The Eildon Hills towering up heather-crowned to the height of over 1,300 feet above the level of the sea right out of the sheen of barley fields, as from a sea of silver, form one of the salient features of this glorious landscape. This is an interesting peculiarity of Scotch scenery;—civilization sapping the barbarism of the wilderness; wheat-fields *mordant* biting in upon peaty moorlands, or climbing to the tops of cold, bald mountains, shearing off their thorny locks of heather and covering them with the well-dressed *chevelure* of yellow grain. Where the farmer's horse cannot climb with the plough, or the little sheep cannot graze to advantage, human hands plant the Scotch larch or fir, just as a tenant-gardener would set out cabbage-plants in odd corners of his little holding which he could have no other use for.

Abbotsferry is just above Abbotsford, and is crossed in a small row-boat. The river here is of considerable width and quite rapid. The boat was kept on the other side; so I hallooed to a man engaged in thatching a rick of oats to come and ferry me over. Without descending from the ladder, he called to some one in the cottage, when, to my surprise, a well-dressed young woman, in rather flowing dress, red jacket, and with her hair tastefully done up in a net *a-la-mode*, made her appearance. Descending to the river, she folded up her gown, and, settling herself to the oars, "pushed her light shallop from the shore" with the grace of The Lady of the Lake. In a few minutes she ran the prow upon the pebbled beach at my feet, and I took my seat at the other end of the boat. She did it all so naturally, and without any other flush upon her pleasant face than that of the exercise of rowing, that I felt quite easy myself and checked the expression of regret I was on the point of uttering for putting her to such service. A few questions convinced me it was her regular employment, especially when her father was busy. I could not help asking her if she had ever read "The Lady of the Lake," but found that neither that romance nor any other had ever invested her river experience with any sensibility except of a cheerful duty. She was going to do the whole for a penny, her usual charge, but I declined to take back any change for the piece of silver I gave to her, intimating that I regarded it cheap at that to be rowed over a river by such hands.

Almost opposite to Abbotsford I passed one of the best farming establishments I had seen in Scotland. I was particularly struck with a feature which will hereafter distinguish the steddings or farm buildings in Great Britain. Steam has already accomplished many changes, and among others one that could hardly have been anticipated when it was first applied to common uses. It has virtually turned the threshing-floor out of doors. Grain growing has become completely out-of-door work, from seeding to sending to market. The day of building two-story barns for storing and threshing wheat, barley and oats is over, I am persuaded, in this country. A quadrangle of slate-roofed cow-sheds, for housing horses and cattle, will displace the old-fashioned barns, each with its rood of roof. This I saw on crossing the Tweed was quite new, and may serve as a model of the housing that will come into vogue rapidly. One familiar with New England in the "old meeting-house" time would call this establishment a hollow square of horse-sheds, without a break or crevice at the angles.

I reached Galashiels about 5 p.m., and stopped an hour for tea. This is a vigorous and thrifty town, that makes a profitable and useful business of the manufacture of tweeds, tartans and shawls. It is situated on the banks of the Gala, a little, rapid, shallow river that joins the Tweed about a mile below. After tea I resumed my walk, but owing to the confused direction of the landlady, took the wrong side of the river, and diverged westward toward Peebles. I had made three miles or more in this direction before I found out my mistake, so was obliged to return to Galashiels, where I concluded to spend the night, after another involuntary excursion more unsatisfactory than my walk around Sheffield, inasmuch as I had to travel over the same road twice for the whole distance. Thus the three mistakes thus far made have cost me twenty miles of extra footing. The next morning I set out in good season, determined to reach Edinburgh, if possible, by night.

Followed the Gala Water, as it is called here, just as if it were a placid lake or land-locked bay, though it is a tortuous and swift-running stream. The scenery was still picturesque, in some places very grand and romantic. There was one great amphitheatre just before reaching the village of Stow which was peculiarly interesting. It was a great bowl full of earth's glory up to the very rim. The circular wall was embossed with the best patterns and colors of vegetation. The hills of every *tournure* showed each in a fir setting, looking, with their sloping fields of grain, like inverted goblets of gold vined with emerald leafwork. In the valley a reaping machine was at work with its peculiar chatter and clatter, and men and women were following in its wake, gathering up and binding the grain as it fell and clearing the way for the next round. Up and down these hills frequently runs a stripe of Scotch firs or larches a few rods wide; here and there they resemble those geometrical figures often seen in gardens and pleasure grounds. The sun peeping out of the clouds, and flooding these features with a sudden, transient river of light, gives them a glow and glory that would delight the artist. After a long walk through such scenery, I reached, late in the evening, *Auld Reekie*, a favorite home-name which the modern Athenians love to give to Edinburgh. Being anxious to push on and complete my journey as soon as practicable, I only remained in the celebrated Scotch metropolis one night, taking staff early next morning, and holding northward towards the Highlands.

Edinburgh has made its mark upon the world and its place among the great centres of the world's civilization. On the whole, no city in Great Britain, or in Christendom, has ever attained to such well-developed, I will not say angular, but salient individuality. This is deep-featured and ineffaceable. It is, not was. Edinburgh has reared great men prolifically and supplied the world with them, and kept always a good number back for itself to give a shaping to others the world needed. Its prestige is great in the production of such intellects. But it keeps up with the times. It is faithful to its antecedents, and appreciates them at their full value and obligation. It does not lie a-bed until noon because it has got its name up for educating brilliant minds. Its grand old University holds its own among the wranglers of learning. Its High School is proportionately as high as ever, notwithstanding the rapid growth of others of the same purpose. Its pulpit boasts of its old mind-power and moral stature. Its Theology stands iron-cabled, grand and solid as an iceberg in the sea of modern speculation, unsoftened under the patter of the heterodox sentimentalities of human philanthropy. It is growing more and more a City of Palaces. And the palaces are all built for housing the poorest of the poor, the weakest of the weak and the vilest of the vile. These hospitals

are the Holyroods of Edinburgh II. They honor it with a renown better than the royal palace of the latter name ever won.

I said Edinburgh the Second. That is correct. There are two towns, the Old and the New; the last about half a century's age. But the oldest will be the youngest fifty years hence. The hand of a "higher civilization," with its spirit-level, pick, plane and trowel, is upon it with the grip of a Samson. That hand will tone down its great distinctive individualities and give it the modern *uniformity* of design, face and feature. All these tall houses, built skyward layer upon layer or flat upon flat, until they show half a dozen stories on one street, and twice that number on the other, are doomed, and they will be done for, one by one in its turn. They probably came in with Queen Mary, and they will go out under the blue-eyed Alexandra. They will be supplanted by the most improved architecture of modern taste and utilitarianism. Edinburgh will be Anglicised and put in the fashionable costume of a progressive age; in the same swallow-tailed coat, figured vest and stovepipe hat worn by London, Liverpool and Manchester. It will not be allowed to wear tweed pantaloons except for one circumstance;—that it is now building its best houses of stone instead of brick.

But there are physical features that will always distinguish Edinburgh from all other cities of the world and which no architectural changes can ever obliterate or deface. There are Arthur's Seat, Salisbury Crags, the Calton Hill, and the Castle Height, and there they will stand forever—the grandest surroundings and garniture of Nature ever given to any capital or centre of the earth's populations.

CHAPTER XVII.

LOCH LEVEN-ITS ISLANDCASTLE— STRATHS—PERTH—SALMON- BREEDING—THOUGHTS ON FISH-

FARMING—DUNKELD—BLAIR ATHOLL—DUCAL TREE-PLANTER— STRATHSPEY AND ITS SCENERY— THE ROADS—SCOTCH CATTLE AND SHEEP—NIGHT IN A WAYSIDE COT- TAGE—ARRIVAL AT INVERNESS.

On Friday, Sept. 11th, I left for the north the morning after my arrival in Edinburgh, hoping to finish my long walk before the rainy season commenced. My old friend and host accompanied me across the Forth, by the Granton Ferry, and walked with me for some distance on the other side; then bidding me God-speed, he returned to the city. The weather was fine, and the farmers were very busy at work. A vast quantity of grain, especially of oats, was cut and ready for carting; but little of it had been ricked in consequence of frequent showers. I noticed that they used a different snath for their scythes here from that common in England. It is in two parts, like the handles of a plough, joining a foot or two above the blade. One is shorter than the other, each having a thole. It is a singular contrivance, but seems to be preferred here to the old English pole. I have never seen yet an American scythe-snath in England or Scotland, although so much of our implemental machinery has been introduced. American manure-forks and hay-forks, axes and augurs you will now find exposed for sale in nearly every considerable town, but one of our beautifully mounted scythes would be a great novelty here.

The scenery varies, but retains the peculiarly Scotch features. Hills which we should call mountains are frequently planted with trees as far up as the soil will lie upon the precipitous sides. On passing one of great height, bald at the top, but bearded to the eyebrows with fir and larch, I asked an elderly man, a blacksmith, standing in his shop-door, if they were a natural growth. He said that he and his two boys planted them all about forty-eight years ago. They were now worth, on an average, twelve English shillings, or about three dollars a-piece.

I lodged in Kinross, a pleasant-faced, quiet and comfortable little town, done up with historical associations of special interest. Here is Loch Leven, serene and placid, like a mirror framed with wooded hills, looking at their faces in it. It is a beautiful sheet of water, taking the history out of it. But putting that in and around it, you see a picture before you that you will remember. Here is more of Mary the Unfortunate. You see reflected in the silver sheen of the lake that face which looks at you with its soft appeal for sympathy in all the galleries of Christendom. Out there, on that little islet, green and low, stands the black castle in which they prisoned her. There they made her trembling, indignant fingers write herself "a queen without a crown." Southward there, where amateurs now fish for trout, young Douglas rowed her ashore with muffled oars so softly that they stirred no ripple at the bow. The keys of the castle they threw into the lake to bar pursuit, lay in the mud for nearly three centuries, when they were found by a lad of the village, and presented to the Earl of Morton, a representative of the Douglas family.

The next day I walked on to Perth, passing through a very interesting section, which nature and history have enriched with landscapes and manscapes manifold. It is truly a romantic region for both these qualities, with delightful views in sudden and frequent alternation. Glens deep, winding and dark, with steep mountain walls folding their tree-hands over the road; lofty hills in full Scotch uniform, in tartan heather and yellow grain plaided in various figures; chippering streams, now hidden, now coming to the light, in white flashing foam in a rocky glade of the dell; straths or savannas, like great prairie gardens, threaded by meandering rivers and studded with wheat in sheaves, shocks and ricks, seen over long reaches of unreapt harvests; villages, hamlets, white cottages nestling in the niches and green gorges of the mountains,—and all these sceneries set in romantic histories dating back to the Danes and their doings in Scotland, make up a prevista for the eye and a revista for the mind that keep both in exhilarating occupation every rod of the distance from Kinross

to Perth.

The road *via* Glenfarg would be a luxury of the first enjoyment to any tourist with an eye to the wild, romantic and picturesque. Debouching from this long, winding, tree-arched dell, you come out upon Strathearn, or the bottom-land of the river Earn, which joins the Tay a few miles below. The term strath is peculiarly a Scottish designation which many American readers may not have fully comprehended, although it is so blended with the history and romance of this country. It is not a valley proper, as we use that term; as the Valley of the Mississippi or the Valley of the Connecticut. If the word were admissible, it might be called most descriptively the land-bay of a river, at a certain distance between its source and mouth, such for instance as the German Flats on the Mohawk, or the Oxbow on the Connecticut, at Wethersfield, in Vermont, or the great onion-growing flat on the same river at Wethersfield in Connecticut. These straths are numerous in Scotland, and constitute the great productive centres of the mountain sections. They are generally cultivated to the highest perfection of agricultural science and economy and are devoted mostly to grain. As they are always walled in by bald-headed mountains and lofty hills, cropped as high as man and horse can climb with a plough and planted with firs and larches beyond, they show beautifully to the eye, and constitute, with these surroundings, the peculiar charm of Scotch scenery. The term is always prefixed to the name of the river, as Strathearn, Strathspey, etc.

I noticed on this day's walk the same singular habit that struck me in the north part of Yorkshire; that is, of cutting inward upon the standing grain. Several persons, frequently women and boys, follow the mowers, and pick up the swath and bind it into sheaves, using no rake at all in the process. So pertinaciously they seem to adhere to this remarkable and awkward custom, that I saw two mowers walk down a hill, a distance of full a hundred rods, with their scythes under their arms, in order to begin a new swath in the same way;

four or five men and women running after them full tilt to bind the grain as it fell! Here was a loss of at least five minutes each to half a dozen hands, amounting to half an hour to a single man at the end of each swath or work. Supposing the mowers made twenty in ten hours from bottom to top of the field, here is the loss of one whole day for one man, or one sixth of the whole aggregate time applied to the harvesting of the crop, given to the mere running down that hill of six pairs of legs for no earthly purpose but to cut inward instead of outward, as we do. The grain-ricks in Scotland are nearly all round and quite small. Every one of them is rounded up at the top and fitted with a Mandarin-looking hat of straw, which sheds the rain well. A good-sized farmhouse is flanked with quite a village of these little round stacks, looking like a comfortable colony of large, yellow tea-caddies in the distance.

Reached Perth a little after dark, having made a walk of nearly twenty miles after 11 a.m. Here I remained over the Sabbath, and greatly enjoyed both its rest and the devotional exercises in some of the churches of the city.

The Fair City of Perth is truly most beautifully situated at the head of navigation on the Tay, as Stirling is on the Forth. It has no mountainous eminence in its midst, castle-crowned, like Stirling, from which to look off upon such a scene as the latter commands. But Nature has erected grand and lofty observatories near by in the Moncrieffe and Kinnoull Hills, from which a splendid prospect is unrolled to the eye. There is some historical or legendary authority for the idea that the Romans contemplated this view from Moncrieffe Hill; and, as the German army, returning homeward from France, shouted with wild enthusiasm, at its first sight, *Der Rhein! Der Rhein*! so these soldiers of the Cæsars shouted at the view of the Tay and the Corse of Gowrie, *Ecce Tiber! Ecce Compus Martius*! There was more patriotism than parity in the comparison. The Italian river is a Rhine in history, but a mere Goose Creek within its actual banks compared with the Tay. In history, Perth has its full share of "love and murder," rhyme and romance, sieges, battering and burning, royals and rebels. In the practical life of to-day, it is a progressive, thriving town, busy, intelligent, respected and honorable. The two natural features which would attract, perhaps, the most special attention of the traveller are the two Inches, North and South, divided by the city. This is a peculiar Scotch term which an untravelled American will hardly understand. It has no relation to measurement of any kind; but signifies what we should call a low, level green or common in or adjoining a town. The Inches of Perth are, to my eye, the finest in Scotland, each having about a mile and a half in circumference, and making delightful and healthy playgrounds and promenades for the whole population.

On Monday, Sept. 14th, I took staff and set out for another week-stage of my walk, or from Perth to Inverness. Crossed the Tay and proceeded northward up the east side of that fertile river. Fertile may sound at first a singular qualification for a broad, rapid stream running down out of the mountains and widening into a bay or firth at its mouth. But it may be applied in the best sense of production to the Tay; and not only that, but other terms known to practical agriculture. Up to the present moment, no river in the world has been cultivated with more science and success. None has been sown so thickly with seed-vitalities or produced more valuable crops of aquatic life. Here salmon are hatched by hand and folded and herded with a shepherd's care. Here pisciculture, or, to use a far better and more euphonious word, fish-farming, is carried to the highest perfection in Great Britain. It is a tillage that must hereafter take its place with agriculture as a great and honored industry. If the cold, bald-headed mountains, the wild, stony reaches of poverty-stricken regions, moor, morass, steppe and prairie are made the pasturage of sheep innumerable, the thousands of rivers in both hemispheres will not be suffered to run to waste through another century. The utilitarian genius of the present age will

turn them into pasturage worth more per acre than the value of the richest land on their banks. Just think of the pasturage of the Tay. It rents for £14,000 a year; and those who hire it must make it produce at least £50,000, or $240,000 annually. Let us assume that the whole length of this salmon-pasturage is fifty miles, and its average width one-eighth of a mile. Then the whole distance would contain the space of 4,000 square acres, and the annual rent for fishing would amount to over £3 13s. per acre. This would make every fish-bearing acre of the river worth £100, calculated on the land basis of interest or rent.

Having heard of the Stormontfields' Ponds for breeding salmon, I had a great desire to see them. They are situated on the Tay, a few miles above Perth, and are well worthy of the inspection and admiration of the scientific as well as the utilitarian world. The process is as simple as it is successful and valuable. A race or canal, filled with a clear, mountain stream, and constructed many years ago to supply motive power to a corn-mill, runs parallel with the river, at the distance from it of about twenty rods. At right angles with this stream, there are twenty-five wooden boxes side by side, about fifty feet in length, placed on a slight decline. These boxes or troughs, each about two feet wide and one foot deep, are divided into partitions by cross-boards, which do not reach, within a few inches, the top of the siding, so that the water shall make a continuous surface the whole length of the trough. Each trough is filled with round river stones or pebbles washed clean, on which the spawn is laid. The water is let out of the mill-race upon these troughs through a wire-cloth filter, covering them about two inches deep above the stones. At the bottom, a lateral channel or race, running at right angles to the troughs, conducts the waste water in a rapid, bubbling stream down into the feeding-pond, which covers the space of about one-fifth of an acre, close to the river, with which it is connected by a narrow race gated also with a wire-cloth, to prevent the little living mites from being carried off before their

time.

This may serve to give the reader some approximate idea of the construction of the fish-fold. The next process is the stocking it with the breeding ewes of the sea and river. The female salmon is caught in the spawning season with a net, and the ova are expressed from her by passing the hand gently down the body, when she is again put into the river to go on her way. The manager told me that they generally reckoned upon a thousand eggs to a pound of the salmon caught. Thus fourteen good-sized fish would stock the twenty-five troughs. When hatched, the little things run down into the race-way, which carries them into the feeding-pond. Here they are fed twice daily, with five pounds of beef's liver pulverised. They remain in this water-yard from April to autumn, when the gate is raised and they are let out into the river. And it is a very singular and interesting fact that those only go which have got their sea-coats on them, or have reached the "smolt" character. The smaller fry remain in the pond until, as it has been said in higher circles of society, their beards are grown, or, in their case, until their scales are grown, to fit them for the rough and tumble of salt-water life.

The growth of the little bull-headed mites, after being turned into the river-pasture, is wonderful—more rapid than that of lambs of the Southdown breed. The keeper had marked some of them, on letting them out, by clipping the dorsal fin. On being caught six or eight months afterward, they weighed from five to seven pounds against half a pound each when sent forth to take care of themselves. The proprietors of the fisheries defray the expense of this breeding establishment, being taxed only twopence in the pound of their rental. This, of course, they get back with large interest and profit from the tenant-farmers of the river. As a proof of the enhanced production of the Tay fisheries under this cultivation the fact will suffice, that they now rent for £14,000 a year against £11,000 under the old system.

Salmon-breeding is doubtless destined to rank with sheep-culture and cattle-culture in the future. The remotest colonies of Great Britain are moving in the matter with vigor and almost enthusiasm. Vessels have been constructed on purpose to convey this fair and mottled stock of British rivers to those of Australia and New Zealand. In France, fish-farming has become a large and lucrative occupation. I hope our own countrymen, who plume themselves on going ahead in utilitarian enterprises, will show the world what they can do in this. Surely our New England men, who claim to lead in American industries and ingenuities, will not suffer half a million acres of river-pasturage to run to waste for another half century, when it would fold and feed millions of salmon. Once they herded in the Connecticut in such multitudes that a special stipulation was inserted in the indentures of apprentices in the vicinity of the river, that they should not be obliged to eat salmon more than a certain number of times in a week. Now, if a salmon is caught between the mouth and source of the river, it is blazoned forth in the newspapers as a very extraordinary and unnatural event. There is no earthly reason why the Connecticut should not breed and supply as great a number of these excellent and beautiful fish as the Tay. Its waters are equally pure and quiet as those of the Scotch river. Every acre of the Connecticut, from the northernmost bridge that spans it in Vermont to its debouchment at Saybrook, might be made productive of as great a value as any onion-garden acre at Wethersfield.

The salmon-shepherd at Stormontfields, having fully explained the labors and duties of his charge, rowed me across the Tay, and I continued my walk highly gratified in having seen one of the new industries which this age is adding to the different cultures provided for the sustentation and comfort of human life. The whole way to Dunkeld was full of interest, nature and history making every mile a scene to delight the eye and exhilarate the mind. The first considerable village I passed through was Stanley, which gives the name to

that old family of British peers known in history by the battle-cry of a badly-pressed sovereign, "On, Stanley, on!" Murthley Castle, the seat of Sir William Stewart, and the beautiful grounds which front and surround it, will excite the admiration of the traveller and pay him well for a moment's pause to peruse its illuminated pages opened to his view. The baronet is regarded as an eccentric man, perhaps chiefly because he has built a splendid Roman Catholic chapel quite near to his mansion and supports a priest of that order mostly for his own spiritual good. Near Dunkeld, Birnam Hill lifts its round, dark, bushy head to the height of over 1,500 feet, grand and grim, as if it wore the bonnet of Macbeth and hid his dagger beneath its tartan cloak of firs. "Birnam Wood," which Shakespeare's genius has made one of the immortals among earthly localities, was the setting of that hill in his day, and perhaps centuries before it. Crossing the Tay by a magnificent bridge, you are in the famous old city and capital of ancient Caledonia, Dunkeld. Here centre some of the richest rivulets of Scotch history, ecclesiastical and military, of church and state, cowl and crown. Walled in here, on the upper waters of the Tay, by dark and heavily-wooded mountains, it was just the place for the earliest monks to select as the site of one of their cloistered communities. The two best saints ever produced by these islands, St. Columba and St. Cuthbert, are said to have been connected with the religious foundations of this little sequestered city. The old cathedral, having been knocked about like other Roman Catholic edifices in the sledge-hammer crusades of the Reformation, was *ruined* very picturesquely, as a tourist, with one of Murray's red-book guides in his hand, would be likely to say. But the choir was rebuilt and fitted up for worship by the late Duke of Atholl at the expense of about £5,000.

Of this duke I must say a few words, for he has left the greenest monument to his memory that a man ever planted over his grave. He did something more and better than roofing the choir of a

ruined cathedral. He roofed a hundred hills and valleys with a larch-and-fir work that will make them as glorious and beautiful as Lebanon forever. One of the most illustrious and eloquent of the Iroquois aristocracy was a chief called Corn-planter. This Duke of Atholl should be named and known for evermore as the great Tree-planter of Christendom. We have already dwelt upon the benefaction that such a man leaves to coming generations. This Scotch nobleman virtually founded a new order of knighthood far more useful and honorable than the Order of the Garter. To talk of *garters*!—why, he not only put the cold, ragged shivering hills of Scotland into garters, but into stockings waist high, and doublets and bonnets and shoes of beautifully green and thick fir-plaid. He planted 11,000 square acres with the larch alone; and thousands of these acres stood up edgewise against mountains and hills so steep that the planters must have spaded the holes with ropes around their waists to keep them from falling down the precipice. It is stated that he had twenty-seven millions of the larch alone planted on his mountainous estates, besides several millions of other trees. Now, it is doubtful if the whole region thus dibbled with this tree-crop yielded an average rental of one English shilling per acre as a pasturage for sheep. On passing through miles and miles of this magnificent wood-grain and taking an estimate of its value, I put it at 10s., or $2 40c. per tree. Of the twenty-seven millions of larches thus planted, ten must be worth that sum; making alone, without counting the rest, £5,000,000, or $24,000,000. It is quite probable that the larches, firs and other trees now covering the Atholl estates, would sell for £10,000,000 if brought to the hammer. But he was not only the greatest arboriculturist in the world, but the founder of tree-farming as a productive industry as well as a decorative art. Already it has transformed the Highlands of Scotland and trebled their value, as well as clothed them with a new and beautiful scenery. What we call the Scotch larch was not originally a native of that country. Close to the cathedral in Dunkeld stand the two patriarchs of the family, first introduced into Scotland from Switzerland in 1737.

Having remained the best part of two days in Dunkeld, I held on northward, through heavily-shaded and winding glen and valley to Blair Atholl. For the whole distance of twenty miles the country is quite Alpine, wild and grand, with mountains larched or firred to the utmost reach and tenure of soil for roots; deep, dark gorges pouring down into the narrowing river their foamy, dashing streams; mansions planted here and there on sloping lawns showing sunnily through groves and parks; now a hamlet of cottages set in the side of a lofty hill, now a larger village opening suddenly upon you at the turning of the turnpike road. I reached Blair Atholl at about dark, and lodged at the largest hotel I slept in between London and John O'Groat's. It is virtually the tourist's inn; for this is the centre of some of the most interesting and striking sceneries and localities in Scotland. Glens, waterfalls, stream, torrent, mountain and valley, with their romantic histories, make this a very attractive region to thousands of summer travellers from England and other countries. The railway from Perth to Inverness via Dunkeld and Blair Atholl, has just opened up this secluded Scotch Switzerland to multitudes who never would have seen it without the help of the Iron Horse. A month previous, this point had been the most distant in Scotland from steam-routes of transportation and travel. Now southern sportsmen were hiring up "the shooting" for many miles on both sides of the line, making the hills and glens echo with their fusillades. Blair Castle, the duke's mansion, is a very ordinary building in appearance, looking from the public road like a large four-story factory painted white, with small, old-fashioned windows. He himself was lying in a very painful and precarious condition, with a cancer in the throat, from which it was the general impression that he never would recover. The day preceding, the Queen had visited him, while *en route* for Balmoral, having

gone sixty miles out of her way to comfort him with such an expression of her sympathy.

The next day I reached the northern boundary of the Duke of Atholl's estates, having walked for full forty miles continuously through it. Passed over a very bleak, treeless, barren waste of mountain and moorland, most of it too rocky or soilless for even heather. The dashing, flashing, little Garry, which I had followed for a day or two, thinned and narrowed down to a noisy brook as I ascended towards its source. For a long distance the country was exceedingly wild and desolate. Terrible must be the condition of a man benighted therein, especially in winter. There were standing beacons all along the road for miles, to indicate the track when it was buried in drifting snow. These were painted posts, about six or eight feet high, planted on the rocky, river side of the road, at a few rods interval, to guide the traveller and keep him from dashing over the concealed precipices. About the middle of the afternoon I reached the summit of the two watersheds, where a horse's hoof might so dam a balancing stream as to send it southward into the Tay or northward into the Moray Firth. Soon a rivulet welled out in the latter direction with a decided current. It was the Spey. A few miles brought me suddenly into a little, glorious world of beauty. The change of theatrical sceneries could hardly have produced a more sudden and striking contrast than this presented to the wild, cold, dark waste through which I had been travelling for a day. It was Strathspey; and I doubt if there is another view in Scotland, of the same dimensions, to equal it. It was indescribably grand and beautiful, if you could blend the meaning of these two commonly-coupled adjectives into one qualification, as you can blend two colors on the easel. To get the full enjoyment of the scene at one draught, you should enter it first from the south, after having travelled for twenty miles without seeing a sheaf of wheat or patch of vegetation tilled by the hand of man. I know nothing in America to compare it with or to

help the American reader to an approximate idea of it. Imagine a land-lake, apparently shut in completely by a circular wall of mountains of every stature, the tallest looking over the shoulders of the lower hills, like grand giants standing in steel helmets and green doublets and gilded corselets, to see the soft and quiet beauty of the valley sleeping under their watch and ward. As the sun-bursts from the strath-skies above darted out of their shifting cloud-walls and flashed a flush of light upon the solemn brows of these majestic apostles of nature one by one, they stood haloed, like the favored saints in Scripture in the overflow of the Transfiguration. It was just the kind of day to make the scene glorious indescribably. The clouds and sky were in the happiest disposition for the brilliant plays and pictures of light and shade, and dissolving views of fascinating splendor succeeded and surpassed each other at a minute's interval. Now, the great land-lake, on whose bosom floated in the sunlight a thousand islands   oat-and-barley-gilded,   and rimmed with the green and purple verdure of the turnip and rutabaga, was all set a-glow by a luminous flood from the opening clouds above. The next moment they closed this disparted seam in their drapery, and opened a side one upon the still, grave faces of the surrounding mountains; and, for a few minutes, the smile went round from one to the other, and the great centurions of the hills looked happy and almost human in the gleam. Then shade's turn came in the play, and it played its part as perfectly as light. It put in the touch of the old Italian masters, giving an everchanging background to all the sublime pictures of the panorama.

I was not alone in the enjoyment of this scenery. For the first time in this Walk I had a companion for a day. A clergyman from near Edinburgh joined me at Kingussie, with whom I shared the luxury of one of the most splendid views to be found in Scotland. Indeed, few minds are so constituted as to prefer to see such natural pictures alone. After a day's walk among these sceneries, we came to the small village of Aviemore in the dusk of the evening. Here we found that the only inn had been closed and turned into a private residence, and that it was doubtful if a bed could be had for love or money in the place. The railway through it to Inverness had just been opened, and the navvies seemed still to constitute the largest portion of the population. Neither of us had eaten any dinner, and we were hungry as well as tired. Seeing a little, low cottage near the railroad, with the sign of something for the public good over the door, we went to it, and found that it had two rooms, one a kind of rough, stone-floored shed, the other an apartment full ten feet square, with two beds in it, which occupied half the entire space. But, small as it was, the good man and woman made the most of it in the way of entertainment, getting up a tea occasionally for persons stopping over in the village at a meal-time, also selling small articles of grocery to the laborers. Everything was brought from a distance, even their bread, bacon and butter. Their stock of these fundamentals was exhausted, so that they could not give us anything with our tea until the arrival of the train from the north, which we all watched with common interest. In the course of half an hour it came, and soon our cabin-landlord brought in a large basket full of the simplest necessaries of life, which we were quite prepared to enjoy as its best luxuries. Soon a wood fire blazed for us in the double-bedded parlor, and the unpainted deal table was spread in the fire-light with a repast we relished with a pleasant appreciation.

My companion was bound northward by the next train in that direction, and was sure to find good quarters for the night; but as there was not an inn for ten miles on the route I was to travel, and as it was now quite night and the road mostly houseless and lonely, I felt some anxiety about my own lodging. But on inquiry I was very glad to find that one of the two beds in the room was unoccupied and at my disposal. So, having accompanied my fellow-traveller to the station and seen him off with mutual good wishes, I returned to the cottage,

and the mistress replenished the fire with a new supply of chips and faggots, and I had two or three hours of rare enjoyment, enhanced by some interesting books I found on a shelf by the window. And this is a fact worthy of note and full of good meaning. You will seldom find a cottage in Scotland, however poor and small, without a shelf of books in it. I retired rather earlier than usual; but before I fell asleep, the two regular lodgers, who occupied the other bed, came in softly, and spoke in a suppressed tone, as if reluctant to awaken me. And here I was much impressed with another fact affiliated with the one I have mentioned—that of praying as well as reading in the Scotch cottage. After a little conversation just above a whisper, the elder of the two—and he not twenty, while the other was apparently only sixteen—first read, with full Scotch accent, one of the hard-rhymed psalms used in the Scotch service. Then, after a short pause, he read with a low, solemn voice a chapter in the Bible. A few minutes of silence succeeded, as if a wordless prayer was going upward upon the still wings of thought, which made no audible beating in their flight. It was very impressive; an incident that I shall ever hold among the most interesting of all I met with on my walk. They were not brothers evidently, but most likely strangers thrown together on the railroad. They doubtless came from different directions, but, from Highlands or Lowlands, they came from Bible-lighted homes, whose "voices of the night" were blended with the breathings of religious life and instruction. Separated from such homes, they had agreed to make this one after the same spiritual pattern, barring the parental presence and teaching.

The next day after breakfast, took leave of my kind cottage hosts, exchanging good wishes for mutual happiness. Went out of the amphitheatre of Strathspey by a gateway into another, surrounded by mountains less lofty and entirely covered with heather. For several miles beyond Carr Bridge I passed over the wildest moorland. The road was marked by posts about ten feet

high, painted white within two feet of the top and black above. These are planted about fifteen rods apart, to guide the traveller in the drifting and blinding snows of winter. The road over this cold, desolate waste exceeded anything I ever saw in America, even in the most fashionable suburbs of New York and Boston. It was as smooth and hard as a cement floor. Here on this treeless wild, I met several men at work trimming the edges of the road by a line, with as much precision and care as if they were laying out an aisle in a flower garden. After a walk of about seventeen miles, I reached Freeburn Inn about the middle of the afternoon, and as it began to rain and to threaten bad weather for walking, I concluded to stop there for the night, and found good quarters.

The rain continued in showers, and I feared I should be unable to reach Inverness to spend the Sabbath. There was a cattle fair at the inn, and a considerable number of farmers and dealers came together notwithstanding the weather. Indeed, there were nearly as many men and boys as animals on the ground. A score or more had come in, each leading or driving a single cow or calf. The cattle generally were evidently of the Gaelic origin and antecedents—little, chubby, scraggy creatures, of all colors, but mostly black, with wide-branching horns longer than their fore-legs. Their hair is long and as coarse as a polar seal's, and they look as if they knew no more of housing against snow, rain and wintry winds, or of a littered bed, than the buffaloes beyond the upper waters of the Missouri. One would be inclined to think they had lived from calfhood on nothing but heather or gorse, and that the prickly fodder had penetrated through their hides and covered them with a growth midway between hair and bristles. They will not average over 350 lbs. when dressed; still they seem to hold their own among other breeds which have attracted so much attention. This is probably because they can browse out a living where the Durham and Devon would starve.

The sheep in this region are chiefly the old Scotch breed, with curling horns and crocked faces and legs, such as are represented in old pictures. The black seems to be spattered upon them, and looks as if the heather would rub it off. The wool is long and coarse, giving them a goat-like appearance. They seem to predominate over any other breed in this part of Scotland, yet not necessarily nor advantageously. A large sheep farmer from England was staying at the inn, with whom I had much conversation on the subject. He said the Cheviots were equally adapted to the Highlands, and thought they would ultimately supplant the black faces. Although he lived in Northumberland, full two hundred miles to the south, he had rented a large sheep-walk, or mountain farm, in the Western Highlands, and had come to this section to buy or hire another tract. He kept about 4,000 sheep, and intended to introduce the Cheviots upon these Scotch holdings, as their bodies were much heavier and their wool worth nearly double that of the old black-faced breed. Sheep are the principal source of wealth in the whole of the North and West of Scotland. I was told that sometimes a flock of 20,000 is owned by one man. The lands on which they are pastured will not rent above one or two English shillings per acre; and a flock even of 1,000 requires a vast range, as may be indicated by the reply of a Scotch farmer to an English one, on being asked by the latter, "How many sheep do you allow to the acre?" "Ah, mon," was the answer, "that's nae the way we count in the Highlands; it's how monie acres to the sheep."

At about two p.m., the showers becoming less frequent, I set out with the hope of reaching Inverness before night. The wind was high, the road muddy, or *dirty*, as the English call that condition; and the rain frequently compelled me to seek shelter in some wayside cottage, or under the fir-trees that were planted in groves at narrow intervals. The walking was heavy and slow in face of the frequent showers, and a strong gale from the north-east; so that I was exceedingly glad to reach an inn within four miles of Inverness, where I promised myself comfortable lodgings for the night. It

was a rather large, but comfortless-looking house, evidently concentrating all its entertainment for travellers in the tap-room. After considerable hesitation, the landlady consented to give me bed and board; and directed "the lassie" to make a fire for me in a large and very respectable room on the second floor. I soon began to feel quite at home by its side. My boots had leaked on the way and my feet were very wet and cold; and it was with a pleasant sense of comfort that I changed stockings, and warmed myself at the ruddy grate, while the storm seemed to increase without. After waiting about an hour for tea, I heard the lassie's heavy footstep on the stairs; a knock—the door opens—now for the tray and the steaming tea-pot, and happy vision of bread, oatcake and Scotch *scones*! Alas! what a falling-off was there from this delicious expectation! The lassie had brought a severe and peremptory message from the master, who had just returned home. And she delivered it commiseratingly but decidedly. She was to tell me from him that there was nothing in the house to set before me; that the fair the day before had eaten out the whole stock of his provisions; in short, that I was to take my staff and walk on to Inverness. It was in vain that I remonstrated, pleaded and urged wet feet, the darkness, the wind and rain. "It is so," said the lassie, "and can't be otherwise." She tried to encourage me to the journey by shortening the distance by half its actual miles, saying it was only two, when it was full four, and they of the longest kind. So I went out into the night in my wet clothes, and put the best face and foot to the head-wind and rain that I could bring to bear against them. Both were strong, beating and drenching; and it was so dark that I could hardly see the road. In the course of half an hour, I made the lassie's two miles, and in another, the whole of the actual distance, and found comfortable quarters in one of the temperance inns of Inverness, reaching it between nine and ten at night. Here I spent a quiet Sabbath, which I greatly enjoyed.

CHAPTER XVIII.

INVERNESS—ROSS-SHIRE—TAIN—DORNOCH—GOLSPIE—PROGRESS OF RAILROADS—THE SUTHERLAND EVICTION—SEA-COAST SCENERY—CAITHNESS—WICK: HERRING FISHERIES—JOHN O'GROAT'S: WALK'S END.

Inverness is an interesting, good-sized town, with an intellectual and pleasing countenance, of somewhat aristocratic and self-complacent expression. It is considered the capital of the Highlands, and wears a decidedly metropolitan air. It is well situated on the Ness, just at its *debouchement* into the Moray Firth,—a river that runs with a Rhine-like current through the town and is spanned with a suspension bridge. It has streets of city-built and city-bred buildings, showing wealth and elegance. Several edifices are in process of erection that will rank with some of the best in Edinburgh and Glasgow. It has a long and pretentious history, reaching back to the Romans, and dashed with the romance of the wild ages of the country. Oliver Cromwell, or Sledgehammer II., Macbeth, Thane of Cawdor, Queen Mary, Prince Charlie, and other historical celebrities, entered their names and doings on the records of this goodly town.

On Monday, Sept. 21st, I set out with a good deal of animation on the last week-stage of my journey, which I was anxious to accomplish as soon as possible, as the weather was becoming unsettled with frequent rain. Reached Invergordon, passing through a most interesting section of country, full of very fertile straths. It was the part of Ross-shire lying on the Moray and Beauly Firths and divided by rivers dashing down through the wooded gorges of the mountains. I saw here some of the most productive land in Scotland. Hundreds of acres were studded with wheat and barley stooks, and about an equal space was covered with standing grain, though so near the month of October.

Plantations, parks, gentlemens' seats, glens deep and grand, fir-clad mountains, villages, hamlets and scattered cottages made up the features of every changing view. Indeed, one travelling for a week between Perth and Inverness comes upon such a region as this with pleasant surprise, as upon an exotic section, imported from another latitude.

The next day I held on northward, though the weather was very unfavorable and the walking heavy and fatiguing. Passed what seemed the bold and ridgy island of Cromarty, so associated with the venerated memory of Hugh Miller. The beating rain drove me frequently to the wayside cottages for shelter; and in every one of them I was received with kind words and pleasant looks. One of these was occupied by an old woman in the regular Scotch cap— a venerable old saint, with her Bible and psalm-book library on her window-sill, and her peat fire burning cheerily. When on leaving I intimated that I was from America, she followed me out into the road, asking me a hundred questions about the country and its condition. She had three sons in Montreal, and felt a mother's interest in the very name America. The cottage was one of a long street of them by the sea-side, and I supposed it was a fishing village; but I learned from her that the people were mostly the evicted tenants of the Duke of Sutherland, who were turned out of his county some thirty years ago to make room for sheep. I made only eleven miles this day on account of the rain, and was glad to find cheery and comfortable quarters in an excellent inn kept by a widow and her three daughters in Tain. Nothing could exceed their kindness and attention, which evidently flowed more from a disposition than from a professional habit of making their guests at home for a pecuniary or business consideration. I reached their house about the middle of the afternoon, cold and wet, after several hours' walk in the rain, and was received as one of the family; the eldest daughter, who had all the grace and intelligence of a cultivated lady, helping me off with my wet overcoat, and even offering to pull off my water-soaked boots—an office no American could accept, and which I gently declined, taking the will for the deed. A large number of Scotch *navvies* were at the inns of the town, making an obstreperous *auroval* in celebration of the monthly pay-day. They had received the day preceding a month's wages, and they were now drinking up their money with the most reckless hilarity; swallowing the pay of five long hours at the pick in a couple of gills of whiskey. How strange that men can work in rain, cold and heat at the shovel for a whole day, then drink up the whole in two hours at the gin-shop! These pickmen pioneers of the Iron Horse, with their worst habits, are yet a kind of John-the-Baptists to the march and mission of civilization, preparing its way in the wilderness, and bringing secluded and isolated populations to its light and intercourse. It is wonderful how they are working their way northward among these bald and thick-set mountains. When I first visited Scotland, in 1846, the only piece of railroad north of the Forth was that between Dundee and Arbroath, hardly an hour long. Now the iron pathways are running in every direction, making grand junctions at points which had never felt the navvy's pick a dozen years ago. Here is one heading towards John O'Groat's, grubbing its way like a mole around the firths, cutting spiral gains into the rock-ribbed hills, bridging the deep and dark gorges, and holding on steadily north-poleward with a brave faith and faculty of patience that moves mountains, or as much of them as blocks its course. The progress is slow, silent, but sure. The world, busy in other doings, does not hear the pick, nor the speech of the powder when it speaks to a huge rock a-straddle the path. The world, even including the shareholders, hears but little, if anything, of the progress of the work for months, perhaps for a year. Then the consummation is announced in the form of an invitation to the public to "assist" at the opening of a railroad through towns and villages that never saw the daylight the locomotive brings in its wake. So it will be here. Some day, in the present decade, there will be an excursion train advertised to run from London to John O'Groat's; and perhaps the lineal descendant of Sigurd, or some other old Norse jarl, will wear the conductor's belt and cap or drive the engine.

The weather was still unsettled, with much wind and rain. Resumed my walk, and at about four miles from Tain, crossed the Dornoch Firth in a sail ferry boat, and at noon reached Dornoch, the capital of Sutherlandshire. This was one of the fourteen cities of Scotland; and its little, chubby cathedral, and the tower of the old bishop's palace still give it a kind of Canterbury air. The Earls of Sutherland for many generations lie interred within the walls of this ancient church. After stopping here for an hour or two for dinner, I continued on to Golspie, the residence of the mighty lord of the manor, or the owner, master and human disposer of this great mountain county of Scotland. It is stated that full four-fifths of it belong to him who now holds the title, and that his other great estates, added to this territory, make him the largest landowner in Great Britain and probably in Europe. Just before reaching Golspie, a lofty, sombre mountain, with its bald head enveloped in the mist, and which I had been two hours apparently in passing, cleared away and revealed its full stature—and more. Towering up from its topmost summit, a tall column lifted a human figure in bronze skyward cloud-high and frequently higher still. I believe the brazen face that thus looks into the pure and holy skies without blushing, is a duplicate of the one worn in human flesh by His Grace, Evictor I., who unpeopled his great county of many thousands of human inhabitants, and made nearly its whole area of 18,000 square miles a sheep-walk. But I will not break the seal of that history. It was full of bitter experience to multitudes. Not for the time being was it joyous, but grievous exceedingly—surpassing endurance to many. But it is all over now. The ship-loads of evicted men and women who looked their last upon Scotland while its mountains and glens were reddened with the flames of their burning cottages, carried away with them a bitter feeling in their hearts which years of better experience did not soften. Not for their good did it seem in the motive of the transaction; but for their good it worked most blessedly. It was a rough transplanting, and the tenderest fibres of human affection broke and bled under the uptearing; but they took root in the Western World, and grew luxuriantly under the light and dew of a happier destiny. It was hard for fathers and mothers who were taking on the frostwork of age upon their brows; but for their children it was the birth of a new life; for them it was the introduction to a future which had a sun in it, rayful and radiant with the beams of hope and promise. Let those who denounce and deplore this harsh unpeopling come and stand upon the cold, bleak summit of one of these Sutherland mountains. Let them bring their compasses, or some other instrument for measuring the angles, sines and cosines of human conditions. Plant your theodolite here; wipe the telescope's eye with your handkerchief; look your keenest in the line of the lineage of these evicted thousands. Steady, now! while the most tranquil light of the future is on the pathway of your eye. This first reach of your vision is the life-track of the fathers and mothers unhoused among these mountains. Look on beyond, over the longer lifeline of their children; then farther still under the horizon of the remotest future to the track of their childrens' children. Can you make an angle of a single degree's subtension in the hereditary conditions of these generations, or a dozen beyond? Can you detect a point of departure by which the second generation would have diverged from the first, or the third from the second, and have attained to a higher life of comfort, intelligence, social and political position had they remained in these mountain cottages, grubbed on their cottage farms, and lived from hand to mouth on stinted rations of oatmeal and potatoes, as their ancestors had done from time immemorial? Can you see among all the hopeful possibilities of Time's tomorrows, any

such change for the better? You can sight no such prospect with your telescope in that direction. Turn it around and sweep the horizon of that other condition into which they were thrust, weeping and wrathful against their will. Follow them across the Atlantic to North America, to their homes in the States and in the Canadas. Measure the angle they made in this transposition, and the latitude and longitude of social and moral life they have reached from this Sutherland point of departure. The sons of the fathers and mothers who had their family nests stirred up so cruelly, and scattered, like those of rooks, from their holdings in the cliffs, gorges and glens of these cold mountains, are now among the most substantial and respected men of the Western World. Some of them to-day are mayors of towns of larger population than the whole county of Sutherland. Some, doubtless, are Members of Congress, representing each a constituency of one hundred thousand persons, and a vast amount of intelligence, wealth and industry. They are merchants, manufacturers, farmers, teachers and preachers, filling all the professions and occupations of the continent. Is not that an angle of promise to your telescope? Is not that a line of divergence which has conducted these evicted populations, at a small distance from this point of departure, into the better latitudes of human experience? The selling of this Scotch Joseph to America was more purely and simply a pecuniary transaction than that recorded in Scripture; for in that the unkind and jealous brothers sold the innocent boy for envy, not for the love of pelf, though the Ishmaelites bought him on speculation. But not for envy was the Sutherland lad sold and shipped to a foreign land, but rather for a contemptuous estimate of his money value. The proprietor-patriarch of the county took to a more quiet and profitable favorite—the sheep, and sent it to feed on a pasture enriched with the ashes of Joseph's cottage. It is to be feared he meant only money; but Providence meant a blessing beyond the measurement of money to the evicted; and what Providence

meant it made for him and his posterity, and they are now enjoying it.

Dunrobin Castle, the grand residence of the Duke of Sutherland, looks off upon the sea at Golspie. It is truly a magnificent edifice, ranking with the first palaces in Christendom. Nearly eight hundred years has it been in building, though, I believe, all that commands admiration for stature and style is the work of the present century. Whatever the Sutherland family may have been in local position and history in past centuries, one of the noblest women that ever ennobled the nobility of Great Britain, has given the name a celebrity and an estimation in America which all who ever wore it before never won for it. The Duchess of Sutherland, the noble and large-hearted sister of Lord Morpeth-Carlisle, has given to the coronet she wore a lustre brighter to the American eye than the light of diadems which have dazzled millions in Europe. When the Fatherhood of God and the Brotherhood of Men shall come to its high place in the hearts of nations as the crown-faith of all their creeds, what this noble woman felt, said and did for the Slave in his bonds shall be mentioned of her by the preachers of that great doctrine in years to come. When the jewels of Humanity's memories shall be made up, she who, as it were, bent down to him in his prison-house and put her jewelled hands to the breaking of his fetters, shall stand, with women of the same sympathy, only next to her who broke her box of ointment on the Saviour's feet.

The next day made a walk to Helmsdale, a distance of about eighteen miles. The weather was favorable, the scenery grand and varied with almost every feature that could give it interest. The finest of roads wound in and out around the mountain headlands, so that alternately I was walking upon a lofty esplanade overlooking the still expanse of the steel-blue sea, then facing inward to the gorges of the grand and solemn hills. Found comfortable quarters in one of the inns of Helmsdale, a vigorous, busy, fishing village nestling under the shadow of the mountains at the mouth of a little river of the same name. After tea, went down to the wharf or quay and had some conversation with one of the masters of the business. He cured and put up about 30,000 barrels of herrings himself in a season, employing, while it lasted, 500 persons. Their chief market is the North of Europe, especially Poland, and the business was consequently much depressed on account of the troubles in that country. The occupation of this little sea-side village illustrated the ramifications of commerce. They imported their salt from Liverpool, their staves from Norway and their hoops from London.

Set out again immediately after breakfast, feeling that I was drawing near to the end of my journey. I was soon in the treeless county of Caithness, so fraught with the wild romance of the Norsemen. Passed over the bleakest district I had yet seen, called Old Ord, a cold, rough, cloud-breeding region that the very heavens above seem to frown upon with a scowl of dissatisfaction. Still, the road over this dark, mountain desert, though staked on each side to keep the traveller from wandering in the blinding snows of winter, was as beautifully kept as the carriage-way in the park of Dunrobin Castle. The sending of an English queen to conciliate the Welsh, by giving birth to a son in one of their castles, was not a much better stroke of policy than that of England in perforating Scotland to the Northern Sea with this unparalleled and splendid road, constructed at first for a military purpose. I heard a man repeat a couplet, probably of unwritten poetry, in popular vogue among the Highlands, and which has quite an Irish collocation of ideas. It is spoken thus, as far as I can recollect—

Who knew these roads ere they were made
Should bless the Lord for General Wade.

I doubt if there are ten consecutive miles of carriage-road in America that could compare for excellence with that over the desert of Old Ord. I was overtaken by a heavy shower before I had made the *trajet*, and was glad to reach

one of the most comfortable inns of the Highlands, in the beautiful, romantic and picturesque glen of Berriedale. Here, nestling between lofty mountain ridges, which warded off the blasting sea-winds sweeping across from Norway, were plantations and groves of trees, almost the only ones I saw in the county. Nothing could exceed the hospitality of the family that kept the large, white-faced hotel at the bottom of this pleasant valley; especially after I incidentally said that I had walked all the way from London to see the country and people. They admitted me into the kitchen and gave me a seat by the great peat fire, where I had a long talk with them, beginning with the mother. Having intimated that I was an American, the whole family, old and young, including the landlord, gathered around me and had a hundred questions to ask. They related many incidents about the great eviction in Sutherland, which was an event that seems to make a large stock of legendary and unwritten stories, like the old *Sagas* of the Northmen. When I had dried my clothes and eaten a comfortable dinner before their kitchen fire and resumed my staff, they all followed me out to the road, and then with their wishes for a good journey as long as I was in hearing distance. Continued my walk around headlands, now looking seaward, now mountainward, now ascending on heather-bound esplanades, now descending in zig-zag directions into deep glens, over massive and elegant bridges that spanned the mountain streams and their steep and jagged banks. After a walk of eighteen miles, put up at an inn a little north of the village of Dunbeath, kept by an intelligent and industrious farmer. The rain had continued most of the day, and I was obliged to seek shelter sometimes under a stunted tree which helped out the protecting power of a weather-beaten umbrella; now in the doorway of an open stable or cow-shed, and once with my back against the door of a wayside church, which kept off the rain in one direction. This being a kind of border-season between summer and autumn, there were no fires in the inns generally

except in the kitchen, and I soon learned to make for that, and always found a kindly welcome to its comforts; though sometimes the good woman and her lassie would look a little flushed at having their busiest culinary operations revealed so suddenly to a stranger. Some of these kitchens are fitted for sleeping apartments; occasionally having two tiers of berths like a ship's cabin, slightly and rudely curtained.

The family of this wayside inn, seemingly like every other family in the country, had connections in America, embracing brothers, uncles and cousins. I was shown a little paper casket of hair flower-work, sent by *post*! It was wrought of locks of every shade and tint, from the snow of a grandmother over one hundred years of age to the little, sunny curls of the youngest child in the circle of kindred families. The Scotch branch had collected specimens from relatives in Great Britain and forwarded them to the family in America, one of whose daughters had worked them into two bouquets of flowers, sending one of them by post to this little, white cottage on the Northern Sea, as a memento of affection. What enhanced the beauty of this interchange was the fact, that forty-eight years had elapsed since the landlord's brother left his native land for New England, and had never seen it since. Still, the cousins, who had never seen each other's faces, had kept up an affectionate correspondence. A son and son-in-law of the brother in America were in the Federal army, and here was a sea-divided family filled with all the sad, silent solicitude of affection for beloved ones exposed to the fearful hazards of a war sundering more ties of blood-relationship than any other ever waged on earth.

Saturday, September 27th. Resumed my walk with increased animation, feeling myself within two days' distance of its end. The scenery softens down to an agricultural aspect, the country declining northerly toward the sea. Passed through a well-cultivated district, never unpeopled or wasted by eviction, but held by a kind of even yeomanry of proprietors. The cottages are comfortable, resembling the white houses of New England considerably. They are nearly all of one story, with a chimney at each end, broadside to the road, and a door in the middle, dividing the house into two apartments. They are built of stone, the newest ones having a slate roof. Some of them are whitewashed, others so liberally jointed with mortar as to give them a bright and cheery appearance. These, of course, are the last edition of cottages, enlarged and amended in every way. The old issues are ragged volumes, mostly bound in turf or bog grass, well corded down with ropes of heather, giving the roof a singular ribby look, rounded on the ridge. In many cases a stone is attached to each end of the rope, so as to make it hug the thatch closely. I noticed that in a considerable number of the old cottages, the stone wall only reached up a foot or two from the ground, the rest being made up of blocks of peat. Some of the oldest had no premonitory symptoms of a chimney, except a hole in the roof for the smoke. These in no way differed from the stone-and-turf cottages in Ireland.

Again occasional showers brought me into acquaintance with the people living near the road. In every case I found them kind and hospitable, giving me a pleasant welcome and the best seat by their peat-fire. I sat by one an hour while the rain fell cold and fast outside. The good woman and her daughter were busy baking barley-cakes. They were the first I had seen, and I ate them with a peculiar zest of appetite. Told them many stories about America in return for a great deal of information about the customs and condition of the working-people. They generally built their own cottages, costing from £40 to £50, not counting their own labor. I met on the road scores of fishermen returning to their homes at the conclusion of the herring season; and was struck with their appearance in every way. They are truly a stalwart race of men, broad-chested, of intelligent physiognomy, with Scandinavian features fully developed. A half dozen of them followed a horse-cart containing their nets, all done up in a round ball, like a bladder of snuff,

with the number of their boat marked upon it.

At about four p.m., I came in sight of the steeples of Wick, a brave little city by the Norse Sea, which may not only be called the Wick but the Candle of Northern Scotland; lighting, like a polar star, this hyperborean shoreland of the British isle. I never entered a town with livelier pleasure. It is virtually the last and farthest on the mainland in this direction. Its history is full of interest. Its great business is full of vigor, daring and danger. Here is the great land-home of the Vikings of the nineteenth century; the indomitable men who walk the roaring and crested billows of this Northern Ocean in their black, tough sea-boats and bring ashore the hard-earned spoils of the deep. This is the great metropolis of Fishdom. Eric the Red, nor any other pre-Columbus navigator of the North American Seas, ever mustered braver crews than these sea-boats carry to their morning beats. Ten thousand of as hardy men as ever wrestled with the waves, and threw them too, are out upon that wide water-wold before the sun looks on it—half of them wearing the features of their Norse lineage, as light-haired and crisp-whiskered as the sailors of Harold the Fair-haired a thousand years ago. They come from all the coasts of Scotland, from Orkney, Shetland, the Hebrides and Lewis islands, and down out of the heart of the Highlands. It is a hard and daring industry they follow, and hundreds of graves on the shore and thousands at the bottom of the sea have been made with no names on them, as the long record of the hazards they run in the perilous occupation. But they keep their ranks full from year to year, pushing out new boats marked with higher numbers.

The harbor has been dangerous and difficult of access, but of late a great effort has been made to render it more safe and commodious. The Scotch fisheries now yield from 600,000 to 700,000 barrels of herrings annually, employing about 17,000 fishermen; Wick stands first among all the fishing ports of the kingdom. It is a thriving town, well supplied with churches,

schools, hotels, banks and printing-offices. Several new buildings are now being erected which will rank high in architecture and add new features of elegance to the place. The population is a vigorous, intelligent, highly moral and well-read community, as I could not fail to notice on attending service on the Sabbath at different places of worship. Wick is honored with this distinction— it assembles a larger congregation of men to listen to the glad Evangel on Sunday than any city of the world ever musters under one roof for the same purpose. It is the out-door church of the fishermen. They sometimes number 5,000 adult men, sea-beaten and sun-burnt, gathered in from mountainous island and mainland all around the northern coasts of Scotland.

Monday, Sept. 28th. The weather was favorable, and I set out on my last day's walk northward with a sense of satisfaction I could hardly describe. The scenery was beautiful in every direction. The road was perfect up to the last rod; as well kept as if it ran through a nobleman's park. The country most of the way was well cultivated—oats being the principal crop. Here, almost within sight of the Orkneys, I heard the clatter of the reaping machine, which, doubtless, puts out the same utterance over and upon the sea at Land's End. It has travelled fast and far since 1851, when it first made its appearance in Europe in the Crystal Palace, as one of the wild, impracticable "notions" of American genius. In Wick I visited a newspaper establishment, and saw in operation one of the old "Columbians," or the American printing-press, surmounted by the eagle of the Republic. The sewing-machine is in all the towns and villages on the island. If there is not an American clock at John O'Groat's, I hope some of my fellow townsmen will send one there, Bristol-built. They are pleasant tokens of free-labor genius. No land tilled by slaves could produce them. I saw many large and highly-cultivated farms on these last miles of my walk. The country was proportionately divided between food and fuel. Oats and barley constitute the grain-crops. The uncultivated land interspersed with the yellow fields of harvest, is reserved for *peat*—the poor man's fuel and his wealth. For, were it not for the inexhaustible abundance of this cheap and accessible firing, he could hardly inhabit this region. It would seem strange to an American, who had not realised the difference of the two climates, to see fields full of reapers on the very threshold of October, as I saw them on this last day's walk. I counted twelve women and two men in one field plying the sickle, all strongly-built and good-looking and well-dressed withal.

The sea was still and blue as a lake. A lark was soaring and warbling over it with as happy and hopeful a voice as if it were singing over the greenest acres of an English meadow. When I had made half of the seventeen miles between Wick and John O'Groat's, I began to look with the liveliest interest for the first glimpse of the Orkneys; but projecting and ragged headlands intercepted the prospect. About three p.m., as the road emerged from behind one of them, those famous islands burst suddenly into view! There they were!—in full sight, so near that their grain-fields and white cottages and all their distinguishing features seemed within half a mile's distance. This was the most interesting *coup d'œil* that I ever caught in any country. Here, then, after weeks and months of travel on foot, I was at the end of my journey. Through all the days of this period I had faced northward, and here was the *Ultima Thule*, the goal and termination of my tour. The road to the sea diverged from the main turnpike, which continued around the coast to Thurso. Followed this branch a couple of miles, when it ended at the door of a little, quiet, one-story inn on the very shore of the Pentland Firth. It was a moment of the liveliest enjoyment to me. When I left London, about the middle of July, I was slowly recovering from a severe indisposition, and hardly expected to be able to make more than a few miles of my projected walk. But I had gathered strength daily, and when I brought up at this little inn at the very jumping-off end of Scotland, I was

fresher and more vigorous on foot than at any previous stage of the journey.

Having found to my great satisfaction that they could give me a bed for the night, I went with two gentlemen of the neighborhood to see the site of the celebrated John O'Groat's House, about a mile and a half from the inn. There was only a footpath to it across intervening fields, and when we reached it, a rather vigorous exercise of the organ of individuality was requisite to "locate" the foundations of "the house that Jack built." Indeed, pilgrims to the shrine of this famous domicile are liable to much disappointment at finding so little remaining of a residence so historical. Literally not one stone is left upon another. A large stone granary standing near is said to have been built of the *debris* of the house, and this helps out one's faith when struggling to believe in the existence of such a building at all. A certain ridgy rising in the ground, to which you try to give an octagonal shape, is pointed out as indicating the foundations; but an unsatisfactory obscurity rests upon the whole history of the establishment. Whether true or not, that history of the house which one would prefer to believe runs thus:—

In the reign of James IV. of Scotland, three brothers, Malcolm, Gavin, and John de Groat, natives of Holland, came to this coast of Caithness, with a letter in Latin from that monarch recommending them to the protection and countenance of his subjects hereabout. They got possession of a large district of land, and in process of time multiplied and prospered until they numbered eight different proprietors by the name of Groat. On one of the annual dinners instituted to commemorate their arrival in Caithness, a dispute arose as to the right of precedency in taking the door and the head of the table. This waxed very serious and threatened to break up these annual gatherings. But the wisdom and virtue of John prevented this rupture. He made a touching speech to them, soothing their angry spirits with an appeal to the common and precious memories of their native land and to all their joint experiences in this. He entreated

them to return to their homes quietly, and he would remedy the current difficulty at the next meeting. Won by his kindly spirit and words, they complied with his request. In the interval, John built a house expressly for the purpose, of an octagonal form, with eight doors and windows. He then placed a table of oak, of the same shape, in the middle; and when the next meeting took place, he desired each head of the different Groat families to enter at his own door and sit at the head of his own table. This happy and ingenious plan restored good feeling and a pleasant footing to the sensitive families, and gave to the good Dutchman's name an interest which it will carry with it forever.

After filling my pockets with some beautiful little shells strewing the site of the building, called "John O'Groat's buckies," I returned to the inn. One of the gentlemen who accompanied me was the tenant of the farm which must have been John's homestead, containing about two hundred acres. It was mostly in oats, still standing, with a good promise of forty bushels to the acre. He resided at Thurso, some twenty miles distant, and found no difficulty in carrying on the estate through a hired foreman. I never passed a more enjoyable evening than in the little, cozy, low-jointed parlor of this sea-side inn. Scotch cakes never had such a relish for me nor a peat-fire more comfortable fellowship of pleasant fancies, as I sat at the tea-table. There was a moaning of winds down the Pentland Firth—a clattering and chattering of window shutters, as if the unrestful spirits of the old Vikings and Norse heroes were walking up and down the scene of their wild histories and gibbering over their feats and fates. Spent an hour or two in writing letters to friends in England and America, to tell them of my arrival at this extreme goal of my walk, and a full hour in poring over the visitors' book, in which there were names from all countries in Christendom, and also impressions and observations in prose, poetry, English, French, Latin, German and other languages. Many of the comments thus recorded intimated some dissatisfaction that John O'Groat's House was so *mythical*; that so much had to be supplied by the imagination; that not even a stone of the foundation remained in its place to assist fancy to erect the building into a positive fact of history. But they all bore full and sometimes fervid testimony to the good cheer of the inn at the hands of the landlady. There was one record which blended loyalty to palate and patriotism—"The Roast Beef of Old England" and "God save the Queen"—rather amusingly. A party wrote their impressions after this manner—"Visited John O'Groat's House; found little to see; came back tired and hungry; walked into a couple of tender chickens and a good piece of bacon: God save Mrs. Manson and all the Royal Family!" This concluding "sentiment" was doubtless sincere and honest, although it involved a question of precedence in the rank of two feelings which John the Dutchman could have hardly settled by his eight-angled plan of adjustment.

The next morning, for the first time for nearly three months of continuous travel, I faced southward, leaving behind me the Orkneys unvisited, though I had a strong desire to see those celebrated islands—the theatre of so much interesting history. Twenty years ago I translated all the "*Sagas*" relating to the voyages and exploits of the Northmen in these northern seas and islands, their explorations of the coast of North America centuries before Columbus was born, their doings in Iceland and on all the islands great and small now forming the British realms. This gave an additional zest to my enjoyment in standing on the shore of the Pentland Firth and looking over upon the scene of old Haco's and Sigurd's doing, daring and dying.

Footed it back to Wick, and there terminated my walk, having measured, step by step, full seven hundred miles since I left London, counting in the divergences from a straight line which I had made. In the evening I addressed a large and intelligent audience which had been convened at short notice, and I never stood up before one with such pe-

culiar satisfaction as in that North-star town of Scotland. I had travelled nearly the whole distance *incog.*, without hearing my own name on a pair of human lips for weeks. To lay aside this embargo and to speak to such a large congregation, face to face, was like coming back again into the great communions of humanity after a long and private fellowship with the secluded quietudes of Nature.

At four p.m. the next day, I took the Thurso coach and passed over in the night the whole distance that had occupied me a week in travelling by staff. Stopped a night in Inverness, another at Elgin, and spent the Sabbath with my friend, Anthony Cruickshank, at Sittyton, about fifteen miles north of Aberdeen.

CHAPTER XIX.

ANTHONY CRUICKSHANK—THE GREATEST HERD OF SHORTHORNS IN THE WORLD—RETURN TO LONDON AND TERMINATION OF MY TOUR.

Sittyton designates hardly a village in Aberdeenshire, but it has become a point of great interest to the agricultural world—a second Babraham. In this quiet, rural district, Anthony Cruickshank, a quiet, modest, meek-voiced member of the Society of Friends, "generally called Quakers," has made a history and a great enterprise of vast value to the world. He is one of those four-handed but one-minded men who, with a pair to each, build up simultaneously two great businesses so symmetrically that you would think they gave their whole intellect, will and genius to one. Anthony Cruickshank, the Quaker of Sittyton, has made but little more noise in the world than Nature makes in building up some of her great and beautiful structures. His footsteps were so light and gentle that few knew that he was running at all, until they saw him lead the racers by a head at the end of the course. The world is wide, and dews of every

temperature fall upon its meadow and pasture lands. Vast regions are fresh and green all the year round, yielding food for cattle seemingly in the best conditions created for their growth and perfection. The highest nobility and gentry of this and other countries are giving to the living statuary of these animals that science, taste and genius which the most enthusiastic artists are giving to the still but speaking statuary of the canvas. The competition in this cultivation of animal life is wide and eager, and spreading fast over Christendom; emperors, kings, princes, dukes and belted barons are on the lists. Antipodean agriculturists meet in the great international *concours* of cattle, horses, sheep and swine. Never was royal blood or the inheritance of a crown threaded through divergent veins to its source with more care and pride than the lineage of these four-footed "princes" and "princesses," "dukes" and "duchesses," and "knights" and "ladies" of the stable and pasture. No peerage ever kept a more jealous heraldry than the herd-book of this great quadruped *noblesse*. The world, by consent, has crowned the Shorthorn Durham as the best blood that ever a horned animal carried in its veins. Princely connoisseurs and amateurs, and all the dilettanti as well as practical agriculturists of Christendom, are giving more thought to the perfection and perpetuation of this blood than to any other name and breed. Still—and this distinction is crowned with double merit by the fact—Anthony Cruickshank, draper of Aberdeen, has worked his way, gradually and noiselessly, to the very head and front of the Shorthorn knighthood of the world. While pursuing the occupation to which he was bred with as much assiduity and success as if it had every thought and activity which a man should give to a business, he built up, at a considerable distance from his warehouse, an enterprise of an entirely different nature, to a magnitude which no other man has ever equalled. He now owns the largest herd of Shorthorns in the world, breeding and feeding them to the highest perfection in the cold and naturally unfertile county of Aberdeen, which no man of less patience and perseverance would select as the ground on which to enter the lists against such an array of competitors in Great Britain and other countries. I regret that my Notes have already expanded to such a volume as to preclude a more extended account of his operations in this great field of usefulness. A few simple facts will suffice to give the reader an approximate idea of what he has done in this department.

About the year 1825, young Cruickshank was put to a Friends' school in Cumberland. He was a farmer's son, and seems to have conceived a great fancy for cattle from childhood. A gentleman resided not far from the school, who was an owner and amateur of Shorthorns, and Anthony would frequently spend his half-holidays with him, inspecting and admiring his herd, and asking him questions about their qualities and his way of treating them. From this school he was sent as an apprentice to a trading establishment in Edinburgh, and at the end of his term set up business for himself as a draper in Aberdeen. All through this period he carried with him his first interest in cattle-culture, but was unable to make a beginning in it until 1837, when he purchased a single Shorthorn cow in the county of Durham, and soon afterward two other animals of the same blood. These constituted the nucleus of his herd at Sittyton. One by one he added other animals of the same stock, purchased in different parts of England, Ireland and Scotland. With these accessions by purchase, and from natural increase, his herd grew rapidly and prospered finely, so that he was obliged to add field to field and farm to farm to produce feed for such a number of mouths. In a few years he reached his present maximum which he does not wish to exceed. That is, his herd now averages annually three hundred head of this noble and beautiful race of animals, or the largest number of them owned by any one man in the world. In 1841, he announced his first sale of young bulls, and every year since that date has put up at public auction the male progeny of the herd. These sales usually take place in the first week of October, and are attended by from 300 to 500 persons from all parts of the kingdom. After carefully inspecting the various lots, they adjourn to a substantial luncheon at twelve o'clock, and at one p.m. they repair to the sale ring and the bidding begins in good earnest, and the auctioneer's hammer falls quick and often, averaging about a minute and a half to each lot. Thus the forty lots of young bulls from six to ten months old are passed away, averaging from 33 to 44 guineas each. Besides these, from fifty to sixty young bulls, cows and heifers are disposed of by private sale during the season, ranging from 50 to 150 guineas, going to buyers from all parts of the world.

It is Mr. Cruickshank's well-matured opinion, resulting from long experience and observation, that there is no breed of cattle so easily maintained in good condition as the Shorthorns. His are fed on pasture grass from the 1st of May to the middle of October, lying in the open field night and day. In the winter they are fed *entirely on oat-straw and turnips*. Not a handful of hay or of meal is given them. The calves are allowed to suck their dams at pleasure. He is convinced that with this simple system of feeding, together with the bracing air of Aberdeenshire, he has obtained a tribe of animals of hardy and robust constitutions, of early maturity, well calculated to improve the general stock of the country.

It was to me a delight to see this, the greatest herd of Shorthorns in the world, numbering animals of apparently the highest perfection to which they could attain under human treatment. What a court and coterie of "princes," "dukes," "knights" and "ladies" those stables contained—creatures that would not have dishonored higher names by wearing them! I was pleased to find that Republics and their less pretentious titles were not excluded from the goodly fellowship of this short-horned aristocracy. There was one grand and noble bull called "President Lincoln," not only, I fancy, out of respect to "Honest Old Abe," but also in reference to the disposition and capacities of the animal. Truly, if let loose in some of our New England fields, he would prove himself a tremendous "railsplitter."

After spending a quiet Sabbath with this old friend and host at his farmhouse at Sittyton, I took the train for Edinburgh and had a week of the liveliest enjoyment in that city, attending the meetings of the Social Science Congress. There I saw and heard for the first time the venerable Lord Brougham, also men and women of less reputation, but of equal heart and will to serve their kind and country. I had intended to make a separate chapter on these meetings and another on the re-unions of the British Association at Newcastle-upon-Tyne, but the space to which this volume must be limited precludes any notice of these most interesting and important gatherings. Stopping at different points on the way, I reached London about the middle of October, having occupied just four months in my northern tour; bringing back a heartful of sunny memories of what I had seen and enjoyed.

***

Lightning Source UK Ltd.
Milton Keynes UK
UKOW07f1833261115

263616UK00010B/260/P